NEW ESSAYS IN FICHTE'S

FOUNDATION
OF THE
ENTIRE DOCTRINE OF
SCIENTIFIC KNOWLEDGE

New Essays in Fichte's

Foundation

of the

Entire Doctrine of

Scientific Knowledge

EDITED BY

Daniel Breazeale

AND

Tom Rockmore

Humanity Books

an imprint of Prometheus Books

59 John Glenn Drive, Amherst, New York 14228-2197

Published 2001 by Humanity Books, an imprint of Prometheus Books

Inquiries should be addressed to
Humanity Books
59 John Glenn Drive
Amherst, New York 14228–2197
VOICE: 716–691–0133, ext. 207
FAX: 716–564–2711
WWW.PROMETHEUSBOOKS.COM

05 04 03 02 01   5 4 3 2 1

Library of Congress Cataloging-in-Publication Data

New essays in Fichte's Foundation of the entire doctrine of scientific knowledge / edited by Daniel Breazeale and Tom Rockmore.
     p. cm.
   Includes bibliographical references.
   ISBN 1–57392–916–X (alk. paper)
   1. Fichte, Johann Gottlieb, 1762–1814. Grundlage der gesamten Wissenschaftslehre. 2. Knowledge, Theory of. I. Breazeale, Daniel. II. Rockmore, Tom, 1942–

B2824 .N49 2001
193—dc21   2001024653

Printed in the United States of America on acid-free paper

CONTENTS

INTRODUCTION

Tom Rockmore

This volume is the third in a series culled from the conferences organized by the North American Fichte Society. It contains a carefully selected series of fresh essays presented at a conference in Shakertown at Pleasantville, Kentucky, May 15–19, 1995, organized around J. G. Fichte's *Foundation of the Entire Wissenschaftslehre* (*Grundlage der gesamten Wissenschaftslehre*, 1794).[1]

Fichte's *Grundlages der gesamten Wissenschaftslehre* (1794) is the first of many versions, some sixteen in all, that he published during his lifetime or that were later found in his *Nachlass*. It is triply interesting as the first major text in the corpus of a philosopher of the first rank, as the first major text of what as the result of Fichte's intervention in the debate was to become post-Kantian German idealism, and as a work brimming with insights that even today retains much of its original interest.

There is probably no typical scenario for the emergence of an important philosophical theory. The positions of the major philosophers arose in very different ways: for Plato through the reporting of the discussions led by his teacher, Socrates; for Kant through a series of writings composed over many years as a result of which he discovered the central insights of his mature theory; for Berkeley as early as his first important text when he was still in his early twenties, and so on. We know that Fichte's position arose when, while still a young man, he was mistakenly identified as the author of Immanuel Kant's long-awaited work on religion, which led to him being awarded a professorship at the University of Jena, at the time the most important German-speaking university.

Fichte's characteristic doctrines were formulated when he was asked to review of book by a little-known contemporary skeptic, G. E. Schulze, who wrote under the pseudonym of Aenesidemus, the Greek skeptic of the Second Academy. The ideas stated in this review provided the basis for Fichte's rapid subsequent formulation of his position for contingent reasons. At that time Fichte began to teach, it was usual for philosophers to use manuals in teaching their classes. For this purpose, Kant used texts prepared by others. G. W. F. Hegel wrote his own. The *Encyclopedia of the Philosophical Sciences* and the *Philosophy of Right* are manuals he composed for his students.

Fichte, who needed such a text for his teaching, hurriedly composed his own from week to week in the form of handouts to students in his classes. When he learned he was to replace K. L. Reinhold in Jena, Fichte had never taught. On Johann Kaspar Lavater's invitation,[2] Fichte gave a series of lectures from February 24 to April 26, 1794, in Zurich before a select audience of about half a dozen pastors and politicians. The rehearsal of the lectures Fichte planned to give in Jena was in fact the original version of the theory, but not its first published version. The first published version of the new position that he was quickly to expound in the *Grundlage der gesamten Wissenschaftslehre* literally emerged page by page as Fichte frantically wrote the manual he thought he required at the beginning of his teaching career.

Great philosophical treatises, those very few books whose ideas continue to resonate throughout the later debate, are sometimes written quickly, even very quickly. It is often said that Kant, who spent a dozen years thinking out his critical philosophy, wrote the *Critique of Pure Reason* much more rapidly, in something approaching six months. We know that the initial version of Fichte's *Wissenschaftslehre* was composed page by page from June 14 until the end of July or the beginning of August 1795. It was originally printed in Leipzig in two installments in 1794 and in 1795. Since Fichte composed his book in great haste, it is not surprising that he was dissatisfied with the published version. Although he remarks in the preface that the book itself was not intended for the public, he allowed the book to reappear in 1802, in the same year that an improved version was also published.

Fichte was disatisfied not only by the various editions of the published version of his position but also by his original position. Although he tried to improve its formulation, in an agonizing series of basic revisions he sought also to rework his position in a way that better corresponded to his conception of it.

Other philosophers, dissatisfied with the results of even the most painstaking efforts over many years, have later rewritten their books. Kant revised the *Critique of Pure Reason* in a second edition. Hegel twice revised the *Encyclopedia*. Yet in comparison, Fichte stands out for his persistent effort to perfect the basic statement of his position in a way that has no known philosophical precedent. From 1794 when the *Wissenschaftslehre* first began to emerge from the press until he died in 1814, Fichte steadfastly and somewhat obsessively attempted no less than fifteen more times to find a satisfactory way to express his central insights.

Striving is a central Fichtean concept. To the best of my knowledge, the history of philosophy records no better instance of a philosopher striving to philosophize. Fichte's remark in passing that he needs half a lifetime of leisure to work out the *Wissenschaftslehre* proved to underestimate the difficulty of the task, which he was never able to complete.[3]

The theory Fichte proposes in the various versions of the *Wissenschaftslehre* is from the beginning determined by two different elements that forever marked his life and thought: his Kantianism, which led him publically to confuse Kant's position with his own; and his equally public enthusiasm for the French Revolution. His Kantianism is a consistent, but easily misunderstood element in his writings from the *Attempt At A Critique of All Revelation (Versuch einer Kritik aller Offenbarung,* 1792), his first important text, which brought him to the attention of his philosophical contemporaries, throughout all his later writings.

Although deeply influenced by the critical philosophy, Fichte's position, which is never just a restatement of Kant's, is always very much his own. Fichte's identification with Kant, which he proclaimed loudly and endlessly, was personally useful, since it enabled an entirely unknown, obviously gifted, but penniless young man at a tender age to acquire an important position in the leading German university of the period. But it was also absurd by any standard, marked by characteristic hyperbole, for instance in his celebrated remark that "the majority of men could sooner be brought to believe themselves a piece of lava in the moon that the take themselves for a *self,*"[4] which he uses to justify the claim that he is the only one to have understood Kant. Fichte even goes so far as to assert that Kant's system is based on his own principles.[5]

Fichte's Kantianism, even in its most extreme moments, is never uncritical but always critical. Fichte's own theory, as Kant soon recognized,[6] is very different from Kant's. Indeed, it could hardly be otherwise, since Fichte, despite his claims to orthodoxy, is never, even at the beginning of his career, a mere disciple of Kant, any more than Friedrich Schelling, who early proclaimed his orthodox Fichteanism, was ever only a mere disciple of Fichte. Despite the proliferation of these and similar claims, from the beginning each was always a major philosopher, revolving in his own philosophical orbit.

Although imbued with the spirit of Kant, Fichte and the other post-Kantian idealists are separated from him by the great French Revolution. The impact of this series of events, still the most important political cataclysm of modern times, continues to reverberate throughout modern society, including modern philosophy. When the Revolution broke out in 1789, Kant, who was sixty-five years old, had nearly finished his life's work. With the exception of the *Critique of Judgment*, which appeared in the next year, all his major writings had already been published. Although he was personally interested in the Revolution, Kant found no place for it in his theories. On the contrary, Fichte and then later Hegel were profoundly marked by the French Revolution, which is directly reflected in their philosophical theories.

The Revolution was linked to the Enlightenment whose main figures were typically concerned to restrain religion and to emancipate reason from faith. It is hence no surprise that, among the great German idealists, only Schelling, who atypically tightens rather than loosens the links between philosophy and religion, was less than enthusiastic about it.

Fichte's interest in the Revolution was deep and deeply linked to his philosophy. The terror refers to the period running from the fall of the Girondins at the end of May 1793 until the fall of Robespierre in July 1794. The difference between the reactions of Fichte and Hegel in this respect are interesting. In 1793, namely during the terror, Fichte, who perceived in the Revolution the advent of political freedom in Europe, published two texts defending the French Revolution and warning against interference with the growth of freedom of thought.[7] He remained concerned with the Revolution in later writings. So in the *Letters to the German Nation* in 1807, he sharply criticized Napoleon, whom he regarded as a counterrevolutionary figure. Ever more moderate, Hegel also thought of the Revolution as marking a historical turning point, and as representing an opportunity for basic changes, although he later thought the moment in which such change was possible belonged to the past. Yet Hegel, who was less enthusiastic about the Revoltuion than Fichte, criticized it sharply in the *Phenomenology* for failing to achieve its goals.

Fichte combines his interest in Kant and his interest in the French Revolution in a striking passage that captures his intent:

My system is the first system of freedom. Just as France has freed man from external shackles, so my system frees him from the fetters of things in themselves, which is to say, from those external influences with which all previous systems—including the Kantian—have more or less fettered man. During the very years when France was using external force to win its political freedom I was engaged in an inner struggle with myself and with all deeply rooted prejudices, and this is the struggle which gave birth to my system. Thus the French nation assisted me in the creation of my system. Its valor encouraged me and gave me the energy I required for grasping my system. Indeed, it was while I was writing about the French Revolution that I was rewarded by the first hints and intimations of this system. Thus in a certain sense this system already belongs to the nation of France. . . .[8]

Fichte's impact on post-Kantian idealism derives from his avowed Kantianism. In a sense, for Kant philosophy begins and ends with the critical philosophy. When he insists on the need for philosophy to be a science, he implies, since there can never be more than one true philosophical system, that philosophy worthy of the name begins with his critical philosophy, which, if it meets this criterion, also brings philosophy to an end.

In Kant's wake, Fichte makes a similar claim about his own system of transcendental philosophy. In remarking that philosophy is not yet a science, he suggests, in opposition to Kant, that the latter's goal has not yet been realized, in a word that it remains unrealized as well by the critical philosophy.[9] Yet in suggesting that even after Kant the task still remaining was to restate the results of the critical philosophy in the form of a scientific system, Fichte sets the agenda that dominates the entire post-Kantian movement in German idealism.

In his effort to correct what he regarded as the prevailing misinterpretation of his position, Kant applied an old theological distinction to textual interpretation in order to distinguish between the theory and the letter of a theory. Beginning with Fichte, the post-Kantians, who were convinced of the importance of Kant's Copernican Revolution in philosophy, but regarded it as an unfinished project, nonetheless thought that it needed to be reformulated according to its spirit, if necessary by disregarding its letter, in order to achieve its intent. Post-Kantian German idealism consists in an effort to carry Kantianism beyond Kant, in a word in an effort to complete Kant's Copernican turn.

Fichte is the key figure in the transition from Kant to post-Kantian German idealism. His influence on this movement derives almost exclusively from the initial version of his *Wissenschaftlslehre*. This treatise made a decisive impression in the debate surrounding Kant's *Critique of Pure Reason*. At a time when many others, including Reinhold and J. S. Beck, routinely claimed to be the only one to understand Kant, Fichte's claim was accepted by the young Schelling and the young Hegel.

In his first published philosophical text, the so-called *Differenzschrift*,[10] Hegel works out the consequences of his view of Fichte's Kantianism, which were to determine the formulation of his own position. Like Fichte, Hegel, who accepts the above-mentioned Kantian idea that there is never more than one true philosophical system, identifies it, as do Kant and Fichte, with the critical philosophy. For Hegel, Fichte has raised the Kantian system to a higher level in his transcendental philosophy. In his account of Fichte, Hegel relies on the initial published version of the *Wissenschaftslehre*. The Fichtean form of Kantianism has been modified by Schelling's philosophy of nature (*Naturphilosophie*). The task confronting anyone who desires to develop further and to complete the Kantian philosophy is to understand the difference between Fichte's and Schelling's versions of the Kantian position.

Although Fichte later revised his *Wissenschaftslehre* many times, the earliest version is the only one that exerted any real influence in post-Kantian German idealism, the only one to which later German idealists responded in their writings. Various reasons can be given for the failure of Fichte's great idealist colleagues to respond to later versions of his position that arguably defuse many of their criticisms of it. In part, this is due to their own rapid development as independent philosophers. In part it is due to the fact that a number of Fichte's revisions of his theory only came to light in his *Nachlaß* after he died. In part it is further due to his ever stormy personal relations with his philosophical colleagues.

Fichte, who left Jena before Hegel arrived, did not have a personal relation to him, although he did have one to Schelling. The friendly early links between Fichte and Schelling rapidly degenerated into hard feelings and even open controversy. Schelling, who initially saw himself as Fichte's devoted disciple, quickly broke with Fichte. The precipitating factor was a quarrel over Schelling's understanding of Fichte's transcendental philosophy. The resultant break was consummated nearly immediately after the appearance of Schelling's *System of Transcendental Idealism*, his first major work, in 1800. In an important letter from Fichte to Schelling, written after receiving Schelling's book, the former complains that his disciple has basically misunderstood him.[11] This letter is quickly followed by Schelling's detailed criticism of Fichte's position, as described in the initial

Wissenschaftslehre, in what is in effect a declaration of philosophical independence, even of philosophical war.[12]

Hegel's break with Fichte occurred in the *Differenzschrift*, in the same text in which he declared his allegiance to Fichte's original position as the highest form of Kantianism. In accepting Schelling's philosophy of nature as a still higher form of Fichte's own Kantianism, he in effect sided with Schelling and against Fichte. Hegel is nothing if not consistent. Slow to make up his mind, he later rarely changes it. The version of Fichte's position that he presents in the *Differenzschrift*, based on the latter's initial version of the *Wissenschaftslehre*, is repeated in almost the same form in a number of later writings.

I will end this introduction with a rapid comment, but no more than that, about the interest of Fichte's theories for the contemporary debate. Fichte, who began to work out his position in the late eighteenth century, died almost two centuries ago. The turning of the historical wheel in the intervening period has wrought profound changes in the most important guidelines for philosophy that held sway during Fichte's lifetime. Not surprisingly, numerous ideas that once seemed indispensable, likely to endure forever, have since lost their hold on philosophy as now understood.

A measure of an important philosopher, such as Fichte, is the ability to continue to speak, often in different ways, to successive generations of students. Yet if his position remains interesting at this late date, other than for its prominent place in the German idealist tradition on the path leading from Kant to Hegel, it is not because of his concern with such characteristic ideas as epistemological foundationalism, transcendental deduction, rigorous system, or a conception of philosophy as science.

Each of these ideas now seems out of date. Cartesian foundationalism, which has been widely influential throughout modern philosophy, no longer seems tenable in any form. Transcendental deduction, a form of epistemological foundationalism pioneered by Kant, has few defenders at present. In Kant's wake, the entire German idealist movement can be regarded as a combined effort uniting philosophers of the first rank to reconstruct the critical philosophy in systematic form, in a word as science. Edmund Husserl still insists on philosophy as systematic science, and Martin Heidegger still insists on philosophy as systematic. Philosophers today are rarely uninterested in either system, or in the idea of philosophy as science. But they are often interested in science, certainly more so than Fichte, who alone among the German idealists was nearly completely ignorant of natural science.

Yet it would be mistaken to conclude that Fichte's work is simply out of date. Many other ideas that were just as important to his position maintain their interest now as when they were originally formulated. Perhaps chief among them is his view of the human subject as basically active, never passive, which is the only assumption to which he was willing to admit in the original formulation of his theory in the 1794 version of the *Wissenschaftslehre*.

Our time, which is influenced by natural science, is characterized by broad "naturalizing" tendencies in a movement away from speculative constructions in philosophy. The French Revolution taught us that what once appeared to be permanent features of reality itself were only transient structures erected by human beings. In the wake of the French Revolution, beginning with Fichte, the German idealists rethought the subject of knowledge and ethics from the perspective of real human beings, who were not situated in the mind of the philosopher but in the social and historical world. This idea is already central to Fichte's nascent theory in the original version of the *Wissenschaftslehre*. I suspect that, in the next century, it is through this conception and its later variants that Fichte will most strongly continue to influence the philosophical debate.

The fourteen texts that appear here fall naturally into five sections. To begin with, Daniel Breazeale provides an overall view of Fichte's *Grundlage der gesamten Wissenschaftslehre*, the topic of this volume. The second section contains accounts of Fichte's understanding of deduction by Steven Hoeltzel and Tom Rockmore. The third section, devoted to special issues in the *Grundlage*, includes studies by Michael Baur on self-measure and self-moderation, by Arnold Farr on reflective judgment and the boundaries of human knowledge, by C. Jeffery Kinlaw on Fichte's theory of time, and by Günter Zöller's study on Fichte's views of positing and determining. Section four comprises comparative studies of special problems in Fichte and Kant by Jere Paul Surber and by Pierre Kerszberg; by Michael Vater on Fichte and Schelling; by Vladimir Zeman on Kant, Fichte, and Hegel; and finally by Curtis Bowman on Jacobi and Fichte. The last section contains accounts of the reactions to the *Grundlage*, including Dale Snow's discussion of its reception in general and George Seidel's discussion of its reception by Hegel.

NOTES

1. *Wissenschaftslehre* (from G. *Wissenschaft* = science + *Lehre* = teaching), which is difficult to render into English, generally refers to Fichte's overall philosophical position.

2. The pastor Johann Kaspar Lavater was a friend of Goethe, and the author of a theory of physiognomy later refuted by Hegel in the *Phenomenology of Spirit*.

3. See Draft of a Letter to Baggesen, April or May 1795, in in *FICHTE: Early Philosophical Writings*, ed. and trans. Daniel Breazeale (Ithaca: Cornell, 1988), p. 376.

4. "Foundations of the Entire Science of Knowledge *(Grundlage der gesamten Wissenschaftlehre),*" in *Fichte: Science of Knowledge (Wissenschaftslehre) with the First and Second Introductions*, ed. and trans. by Peter Heath and John Lachs (New York: Appleton-Century-Crofts, 1970), p. 162.

5. See "To Reinhold, March 1, 1794," in *FICHTE: Early Philosophical Writings*, p. 376.

6. For Kant's critique of his younger colleague, see his "Open letter on Fichte's Wissenschaftslehre, August 7, 1799," in Immanuel Kant, *Philosophical Correspondence, 1759–99*, trans. by Arnulf Zweig (Chicago: University of Chicago Press, 1967), pp. 253–54.

7. See "Zurückförderung der Denkfreiheit" (1793), in J. G. Fichte, *Fichtes Werke* (Berlin : Walter de Gruyter, 1971), vol. 6, pp. 1–36; and "Beitrag zur Berichtigung der Urtheile des Publicums über die französische Revolution" (1793), in *Fichtes Werke*, vol. 6, pp. 37–288.

8. Draft of a letter to Baggesen, April or May 1795, in *FICHTE: Early Philosophical Writings*, pp. 385–86.

9. See "To Reinhold, January 15, 1794," in *FICHTE: Early Philosophical Writings*, p. 372.

10. See G. W. F. Hegel, *The Difference Between Fichte's and Hegel's System of Philosophy*, ed. and trans. H. S. Harris and Walter Cerf (Albany: State University of New York Press, 1977).

11. Fichte's letter to Schelling, dated November 15, 1800, in *Fichte-Schelling Briefwechsel*, ed. Walter Schulz (Frankfurt a. M.: Suhrkamp, 1968), pp. 103–106.

12. See Schelling's letter to Fichte, dated November 19, 1800, in *Fichte-Schelling Briefwechsel*, pp. 107–113.

AN OVERVIEW OF FICHTE'S *GRUNDLAGE*

Inference, Intuition, and Imagination
On the Methodology and Method
of the First Jena Wissenschaftslehre

Daniel Breazeale

Shortly after his departure from Jena, Fichte speculated that the reason hardly anyone seemed to have understood his philosophy was "because no one yet knows *what I am trying to accomplish*."[1] Convinced that this two-hundred-year-old complaint still remains all too valid, I have made several recent efforts to explicate the various "tasks"—both "practical" and "theoretical"—of the Jena *Wissenschaftslehre*. Among the philosophical tasks of *Wissenschaftslehre* is that of *explaining* or *deducing* ordinary experience (i.e., our consciousness of a system of "representations accompanied by a feeling of necessity") while simultaneously affirming human freedom, as well as the task of providing a complete description—or "systematic history"—of those necessary acts by means of which the I posits for itself both its own freedom and the objectivity of the world. These two tasks are, of course, intimately related, since, according to Fichte, it is only by means of the latter that the former can be accomplished. It is further evident that neither can be accomplished from *within* the everyday or natural standpoint of ordinary consciousness, and that *if* a philosophical "deduction" of the latter is possible at all, it will have to be conducted from an artificially constructed standpoint that is in some sense "higher" than the standpoint of what is supposed to be explained thereby.[2]

No sooner does one answer the question concerning *what* philosophy is supposed to accomplish, however, than one faces a new and far more difficult question: namely, precisely *how* does the philosopher propose to accomplish his self-assigned tasks? Even granted that he can somehow elevate himself to the requisite standpoint, what precisely is it that he is *sup-*

posed to be doing once he has arrived there? Another way to ask this same question is, What *method* or *methods* does philosophy employ in "explaining" ordinary consciousness and in "describing" the originary acts of the mind, and how do such methods differ from the familiar methods of, on the one hand, logical inference and conceptual analysis and, on the other, introspective psychology?

Ironically enough, the difficulty of answering such questions about the method of the *Wissenschaftslehre* often seems compounded rather than facilitated by Fichte's own specifically "methodological" or "critical" treatises and remarks. Convinced that Kant had "philosophized far too little about his own philosophizing,"[3] Fichte seems to have been determined not to commit the same error and missed few opportunities to comment upon the special requirements and methods of his own philosophizing. Even during the eight years comprising what I am calling the "period of the Jena *Wissenschaftslehre*"—a period that actually begins in Zurich in late 1793 and ends in Berlin early in 1801—he published no less than three full-scale methodological treatises: *Ueber den Begriff der Wissenschaftslehre* (1794), the two "Introductions" to the unfinished *Versuch einer neuen Darstellung der Wissenschaftslehre* (1797), and the *Sonnenklarer Bericht* (1801). The various philosophical methods prescribed in these works and in the many passing methodological remarks contained in other writings of the period often, however, seem *at variance with one another*. It is, for example, far from clear how the emphasis upon "derivation from a single principle" that one finds in *Ueber den Begriff* is to be reconciled with the appeal to "intellectual intuition" that one finds in the "Introductions" or with the method of quasi-mathematical construction elaborated in the *Sonnenklarer Bericht*. Moreover, it is often difficult to see how the *methodology* presented in these treatises corresponds to the *method* actually employed in such systematic works as the *Grundlage der gesamten Wissenschaftslehre*.

What follows is an attempt to clarify both the declared methodology of the Jena *Wissenschaftslehre* and the relationship between the latter and the actual method of the same. But it is only a first step, since I shall strictly limit my investigation to what Luigi Pareyson and others[4] have described as the "first exposition" of the Jena system, which I understand to include everything written between the fall of 1793 and the end of 1795—that is to say, prior to the earliest attempts at a revised, second exposition of the system in the lectures on *Wissenschaftslehre nova methodo*. In this paper, therefore, I will be mainly concerned with *Ueber den Begriff* and the *Grundlage* of 1794–94. By limiting the scope of my inquiry in this manner I will be able to focus upon the methodological questions that most interest me without having to deal at the same time with the difficult question concerning the apparent *development* of Fichte's thought during over the course of his Jena period.

Even with this self-imposed limitation, it is still no simple matter to determine precisely *how* Fichte thought philosophy should proceed and precisely *what it is* that philosophers are supposed to be doing when they ply their exotic trade. As we shall soon see, even the texts of 1794–95 contain what at least *seem* to be a number of *incompatible* accounts of the method of transcendental philosophy, and it is by no means obvious how these various descriptions are to be reconciled. In examining the methodology and method of the first Jena *Wissenschaftslehre*, therefore, I shall proceed as follows: First, I will consider the explicit philosophical methodology outlined by Fichte in *Ueber den Begriff* and other writings of this period, including the invaluable 1794 lectures "Ueber den Unterschied des Geistes und des Buchstabens in der Philosophie." Then I will describe the actual philosophical method—or better, methods—employed in the two systematic treatises published during this same period, the *Grundlage* itself and the *Grundriß des Eigenthümlichen der Wissenschaftslehre in Rücksicht auf das theoretische Vermögen*. Finally, I will compare the methodology prescribed in the former works with the method employed in the latter and will proved an overall description of what I shall call the "mixed method" of the first Jena *Wissenschaftslehre*, to which I will append a few comments concerning the advantages and problems of such a method.

I. THE DECLARED METHODOLOGY OF THE FIRST JENA *WISSENSCHAFTSLEHRE*

A. Logical inference and conceptual analysis

In describing philosophy as a systematic science in which "all its propositions are joined together in a single first principle in which they unite to form a whole."[5] Fichte seems to imply that the proper method of science is simply that of logical inference (*Folgerung*), which is precisely the term he employs in sections 1 through 4 of *Ueber den Begriff* to describe how the truth of one philosophical propostion is "transferred" to the next in accordance with universal "laws of thinking."[6] This characterization of philosophy as a kind of conceptual or logical *analysis* is repeated in the well-known letter to F. V. Reinhard of January 15, 1794, where Fichte follows Kant in contrasting the methods of mathematics and of philosophy.[7] Whereas mathematics is "able to *construct* its concepts in *intuition*," writes Fichte, "philosophy can and should employ *thinking* in order to *deduce* its concepts from one single first principle," and "the form of the deduction is the same as in mathematics, that is, the form prescribed by general logic." What is lacking from this remark, however, is any apparent appreciation of the important *stricture* that Kant had placed upon philosophical knowledge: namely, that one cannot infer synthetic a priori principles directly from pure concepts, but only indirectly, with reference to possible experience. The philosophical method described in the first half of *Ueber den Begriff* thus seems rather more Cartesian—or rather, Reinholdian—than Kantian. Indeed, were this the sum and substance of Fichte's methodology, the *Wissenschaftslehre* would surely represent a step backward rather than forward in the development of transcendental philosophy.

B. Observation and inner intuition

For content or "truth" to be transferable from one proposition to another in the manner specified, it must first be contained in the *Grundsatz* from which the chain of inferences proceeds. Since such a first principle cannot itself be derived from anything higher, its truth must be unprovable;[8] it must be "true in itself," or self-evidently certain. But how is one supposed to *discover* and to *recognize* such a principle? The philosopher has to recognize the truth of his first principle *immediately*; i.e., he has to *intuit* (*anschauen*) or observe (*beobachten*) it, for only in this way is an epistemic subject immediately related to its object.

Hence Fichte's second point concerning philosophical methodology, a point stressed in section 7 of *Ueber den Begriff*, where the philosopher is characterized as an "observer" of a peculiar sort of object, namely "the system of human knowledge," or "the human mind's modes of acting."[9] Philosophy's reliance upon the method of observation is thus not limited simply to the philosopher's original acquaintance with his starting part or first principle. On the contrary, if philosophy is to be a "real science" (*reele* Wissenschaft) with real content,[10] then the philosopher cannot be content simply to infer one proposition from another, but must also, as it were, "keep an eye" on the "object" designated by these same propositions. He must thus begin with "a first principle that is not simply formal but also real,"[11] and must then observe the content in question. Mere logical inference, however, is insufficient for this purpose, for which some sort of intuition is clearly required. The philosopher has to observe "acts of the mind" to which at least some of his propositions are supposed to *correspond*,[12] for philosophy itself, according to Fichte, is nothing less than a *Darstellung* or "presentation" of something that exists independently of the philosophizing subject: namely, the manner in which every human mind (including that of philosophizing subject) necessarily operates. It is precisely to this "observational" component of the *Wissenschaftslehre* that Fichte was referring when he wrote, in a letter to K. L. Reinhold of July 2, 1795, that "what I am trying to communicate is something that can be neither *said* nor *grasped conceptually*; it can only be *intuited*. My words are only supposed to guide the reader in such a way that the desired intuition is formed within him. I advise anyone who wishes to study my writings to let words be words and simply try to enter into my series of intuitions at one point or another." Thinking about the inferences through which the propositions of the *Wissenschaftslehre* are inferred is thus not enough; for, as Fichte explained to his own students, one cannot understand a single one of these formulas "so long as one does not have the intutuition to which the formula in question refers."[13]

But if the "series of intuitions" obtained by the philosopher are to constitute any sort of *explanation* of original consciousness, they must relate to one another in a quasi-logical manner as ground and consequent. The philosopher therefore not only must oberserve the way the mind acts, but must do so in a manner that permits him to describe how each successive act is conditioned by and in turn conditions other acts. In other words, he must provide a *genetic* description[14] of the acts in question. This is also the meaning of Fichte's frequently misunderstood assertion that philosophers are "historians of the human mind—not, of course, journalists, but rather, writers of pragmatic history."[15] As Fichte's own notes make plain, he employed the phrase "pragmatische Geschichte" (a phrase he appropriated from Ernst Platner) to designate a *critical* and *systematic* account of the *genesis* of something—in this case, of the originary acts of the human mind and hence of the system of experience that is its product.[16] "Pragmatic history," therefore, is just another name for "transcendental deduction."

C. Reflection and imagination

Inference and observation are combined in the charactestic act of the philosopher, that is, the "act of reflection."[17] Fichte's description of philosophy itself as a free product of "the power of freely reflective judgment" ("die reflektierende Urteilskraft . . . in ihrer Freiheit") is clearly influenced by his reading of Kant's third *Critique*, with its sharp contrast between the "determinative" and the "reflective" powers of judgment.[18] Whereas the former operates under universal laws of the understanding and is thus subsumptive or criteriological in character, the latter proceeds from particulars to the universal—not via inductive inference, of course, but rather via a process of imaginative construction.[19]

Fichte, however, goes beyond Kant in treating as products of this power of reflective judgment not just judgments concerning aesthetic taste and natural purposiveness, but also *the judgments of philosophy itself.* What precisely is involved in philosophical reflection? First of all, Fichte always associates philosophical reflection with an act of *abstraction,*[20] and, more specifically, with that act through which the philosopher raises himself from the ordinary to the transcendental standpoint. Such an act is plainly an act of *thinking* and not of intuiting. In this first act of philosophical reflection one voluntarily turns one's attention *away* from objects of experience and reflects instead upon one's own consciousness of these same objects, and only thereby does one finally obtain that concept of pure *Ichheit* or "I-hood," which will then serve as the starting point for a further and very different series of reflections, a series also governed by the laws logical inference.[21] Philosophical reflection is manifestly a kind of thinking.

Yet there would be no point in engaging in such an act of reflective abstraction unless one carefully *paid attention* to the results of the same and thereby obtained an "inner intuition" thereof. Such "attentiveness" (*Aufmerksamkeit*) is thus described by Fichte as an essential ingredient in every act of philosophical reflection.[22] The kind of self-intuition that must accompany reflection is not, however, some sort of superhuman power bestowed only upon artistic and philosophical "geniuses," but is simply the power to *pay attention* to the products of one's own thinking and imagination, a power that anyone can acquire simply by "accustoming himself to disciplined, strict abstraction and by engaging in higher and higher levels of reflection."[23]

Of course, no one can be *required* to develop his reflective capacity in this manner, just as no one can be *required* to philosophize in the first place. Philosophical reflection always begins with a *free act* and continues only as *series* of free acts, and the latter series must always be distinguished carefully both from the necessary, originary acts of the mind that philosophy is trying to describe and from the acts and facts of ordinary consciousness that are the products of the latter. Insofar as he simply *attends to* or *observes* these originary acts, the philosopher might seem to be a passive epistemic subject. But this is a misleading appearance, since the "acts of the mind" that he is supposed to be describing are *never* present as such within ordinary consciousness and are first *raised to consciousness*—and hence produced as representations—only by the act of philosophical reflection itself.[24]

Though philosophical reflection may never *violate* the rules of logical inference nor *dispense with* intuitive confirmation, there is more to such reflection than simply *thinking* and *intuiting*. *Einbildungskraft*, the power of imgaination, is also required. First of all, it is required of the philosopher in the sense that without it he cannot freely propose explanatory hypotheses and solutions to specific problems. No alogorithm determines the *direction* of the philosopher's thoughts and inferences, for this depends upon the specific hypotheses and proof strategies he has decided to explore and to test. In this sense, philosophers are *Versuchern*, forever launching trial balloons, and philosophical reflection is always *experimental* in character.[25] Philosophizing therefore can never be done simply "by the numbers" or "by the book": a certain amount of creative guesswork is always involved—a requirement that Fichte describes variously as "an obscure feeling for truth" or "philosophical genius."[26] Indeed, philosophy itself is an experiment, since we can never know in advance whether the proposed "deduction of experience" will turn out to be possible at all. That a scientific system such as the *Wissenschaftslehre* is actually possible is, to begin with a mere *hypothesis*. Whether it is anything more than this, is something that, in Fichte's words, "depends upon the experiment [*kommt auf den Versuch an*]," for "the possibility of the required science can be established only by the reality of the same."[27]

There is also a second, and equally fundamental, sense in which imagination is required for philosophical reflection. As we already noted, in every act of reflection the attention of the philosopher is simultaneously *directed away from* certain objects (from which he abstracts) and also *turned toward* another (upon which he reflects), and it is only in this manner that the latter become intentional objects of philosophy at all. Though philosophical reflection is always "an act of representing," it is unlike ordinary representational consciousness in that the objects in question in this case are not merely *given to* philosophical reflection, but are also *products of* the same, which is precisely why Fichte would eventually begin to characterize the method of philosophy as one of "*construction*" and emphasize rather than minimize the similarities rather than the differences between the methods of philosophy and of mathematics.[28]

In order to explain how intuition and thought can be intimately connected within a single act of reflection, Fichte appeals to the third essential ingredient in philosophical reflection: namely, *produktive Einbildungskraft*, which, in his lectures on *Geist und Buchstabe in der Philosophie*, he describes as "spirit in the higher sense," understood as the ability to become consciousness not just of the products of our own mental acts but of these very acts themselves.[29] In order for us to be able to observe and to describe such acts, however, we must first "think away" everything else and then convert what is left into intuitiable representations or images (*Bilder*). This creative transformation of abstract thoughts into intuitable images—by virtue of which alone philosophy can become what, according to Fichte, it strives to be: namely, "an accurate schema of the human mind as such"[30]—can be accomplished only by the power of productive imagination (*Einbildungskraft* = power to form images).

The method of philosophy is described by Fichte as a kind of action in which the intellect directs its attention back upon its own acts and then observes and describes them. Such an action "is called *reflection*,"[31] and it cannot be understood as an act of mere thinking or intuiting. Philosophy is supposed "to expand the sphere of [the philosopher's] consciousness into a new domain" and "usher us into a new and higher world,"[32] but this "new domain" or "higher world" is not anything that exists apart from the philosopher's own act of reflection. It is, instead, a product of his own reflective activity—and, more specifically, of his creative imagination. And it is just this, admitted Fichte, "that makes my philosophy difficult, because what is essential therein cannot be approached by means of the understanding, but only by the power of imagination."[33]

Once again, however, the imaginative power required for philospohical reflection is not some sort of divine or innate gift, but is something that anyone can develop simply by exercising and cultivating the powers he already possesses. A particularly effective way to cultivate the power of philosophical reflection, according to Fichte, is to study geometry, because this provides "the best cultivation of the power of imagination under the rule of reason." The student of geometry learns how to employ his imagination to produce mental images, which he grasps by his power of inner intuition and from which he then proceeds to draw universally valid conclusions[34]—a skill that will stand him in good stead when he progresses to philosophy.

II. THE ACTUAL METHOD OF THE GRUNDLAGE AND GRUNDRIß

The *Grundlage* begins with a *task* and an *admission:* "We have to *seek out* the absolutely first and simply unconditioned principle of all human knowledge, a principle that can be neither *proven* nor *determined.*"[35] We do this by abstracting and reflecting until we have, first of all, discovered our first principle—or rather, principles—and then discovered how the principles in questions can be "thought together" without detriment to the originally postulated unity of consciousness. Parts 2 and 3 thus each begin by setting reflection the specific task of finding some way to think without contradiction the first principle of the theoretical and practical portions of the *Grundlage.* In neither case can this task be accompished simply by *analyzing* or *drawing inferences* from these principles. Instead, we must, according to Fichte, "make an experiment" and try out all the various ways of thinking the relationship between the limited I and the limited Not-I, until "after having separated and set aside everything impossible and contradictory, the sought-for only possible way of thinking this relationship has been found."[36] Such an experiment in thinking proceeds according to what Fichte calls a "heuristic method" for selecting *which* concepts need to be analyzed[37] and for actively "seeking out"[38] new synthetic unities. Such a method clearly involves both imagination and inference.[39]

Once the synthesis in question has been successfully "sought out" via dialectical inference, however, the method of philosophical reflection seems to undergo a fundamental shift. In part 2 of the *Grundlage,* for example, we finally recognize that the opposition between the limited I and the limited Not-I can be thought if and only if the I itself is originally limited. But this in turn is thinkable if and only if we admit the necessity of the *Anstoß* or "check" as well as the necessity of ascribing to the mind itself the spontaneous power of "oscillating" (*Schweben*) between infinitude and finitude, thereby making it possible for the I itself to posit for itself and ultimately to become conscious of its own limitations. The power in question is, of course, that of productive imagination; and since the necessity of ascribing such a power to the I has now been logically inferred by means of an imaginative, dialectical analysis of the very concept of the I, the power of productive imagination can now be said to have been established as "a fact [*Faktum*] that is originally present in our mind."[40]

At this point philosophical reflection ceases to be primarily a matter of inference and becomes instead one of intuition,[41] and the philosopher becomes an "observer" of the "original fact" he has just deduced and of the manner in which the latter is posited for itself by the I he is now supposed to be observing.[42] This, presumably, is why Fichte noted in the preface to the *Grundlage* that this is a text that "presupposes that one posses the free power of inner intuition [*das Vermögen der Freiheit der innern Anschauung*]."[43] Furthermore, since what is supposed to be intuited or observed in this case is nothing less than the manner in which the I originally generates its own experience by positing for itself its own original limitation, the philosopher's discursive account of this process can be characterized as a "*genetic*" description or explanation of the latter[44] and thus as "a pragmatic history of the human mind."[45]

Straighforward as all of this may sound, there is nevertheless something very peculiar about the "descriptions" contained in the *Grundlage* and *Grundriß*, for they are supposed to constitute an a priori and *universally valid* account of subjectivity as such. Moreover, the object described in this case is explicitly conceded to have no existence on its own and apart from the philosopher's activitiy of positing it through reflection. Neither the "Begriff des Ich" with which the *Grundlage* commences, nor the "Idee des Ich" with which it concludes, nor any of the "necessary acts of the mind" that fall between these two extremes *ever* appear as such within ordinary consciousness, though they "may appear indirectly within philosophical reflection."[46] Unlike the "given" facts of empirical consciousness, these "necessary facts" of philosophical reflection seem to depend for their "necessity"—*and hence for their very facticity*—upon the process of reflection through which they are "discovered." This important point is summarized as follows at the beginning of the *Grundriß*: "The actions through which the I posits anything at all within itself are here 'facts,' because they are reflected upon. . . . But it does not follow from this that that they are are what we usually call 'facts of consciousness' or that one actually becomes concsious of such facts 'facts of (inner) experience.' . . . One should not consider it an objection to the *Wissenschaftslehre* that it establishes facts that are not to be found within (inner) experience. It does not claim to do this. Instead, it simply demonstrates that something corresponding to a certain thought is present in the human mind."[47]

Without such reflectively established "facts" the *Wissenschaftslehre* would be just one more empty *FormularPhilosophie* and not the "system of real thinking" that it claims to be.[48] And yet it is *thinking* not *intuition* that first establishes "*that* something is a fact,"[49] and, as Fichte explains in his Spirit and Letter lectures, it is the task of the productive imagination to raise these "deduced" facts to the level consciousness and thus to make them intuitable within philosophical reflection. Only then can the philosopher become what he is supposed to be: the observer or pragmatic historian of the I's original activity. The power of productive or creative imagination thus proves to be no less necessary for the descriptive series of observations that we find at the end of part 2 and in almost all of part 3 of the *Grundlage* than it was for the dialectical series of inferences that we encountered in the first portion of part 2. Here too the philosopher may be described as "making an experiment"—though this time an "experiment in *seeing*" rather than an experiment in *thinking*.[50]

Productive imagination is therefore just as essential to the reflective positing of the *Grundlage* and the *Grundriß* as is the employement of logical thinking and inner intuition. Indeed, it appears to be the most essential ingredient of all, for, as Fichte remarks in the end of section 5 of the *Grundlage*: "The *Wissenschaftslehre* is not the sort of thing that can be communicated through the mere letter; instead, it can only be communicated through the spirit; for its basic ideas must be elicited in anyone who studies it by the creative imaginative itself. Nor could this be otherwise in the case of a science that penetrates to the ultimate grounds of human knowledge, for the entire enterprise of the human mind issues from the imagination, and the power of imagination can be grasped only by the power of imagination."[51]

III. CONCLUSIONS

It is clear from the preceding survey that the actual method employed in the systematic treatises of 1794–95 is indeed the same as that presented in the methodological writings of the same period: namely, the method of philosophical reflection. To say that the *method* accords with the *methodology*, however, is not to claim that the method in question is particularly easy to grasp or to apply. Indeed, as I have tried to show in this paper, the reflective method of the early Jena *Wissenschaftslehre* is not in in fact any *single* method at all, but is instead a "mixed method," a deliberate and exquisitely balanced *combination* of the several *different* methods of *logical inference, observation-based description*, and *imaginative construction*.

The best way to appreciate the considerable virtues of such a mixed method is to consider the limitations and dangers of any *one* of these ingredients taken by itself. Though logical inference and rigorous analysis are certainly necessary for any systematic philosophy, they are not, by themselves, enough. On the contrary, mere systematic thinking, unsupplemented by some sort of "intuitive" access to content, will never produce anything but an empty formal system. Like Kant, Fichte was deeply convinced that that it is only through *intuition* that knowledge can obtain any real content, and he refused to exempt philosophical knowledge itself from this requirement. (This was the central point of Fichte's rejection of C. G. Bardili's philosophy of "thinking qua thinking" in his 1800 review of Bardili's *Grundriss der ersten Logic*.)[52]

A philosophy of mere intuition, on the other hand, is no better—indeed it is even worse—than a system of pure thinking. The most common result of the decision to base philosophy upon a method of sheer intuition is the sort of "commonsense philosophy" that appeals to the purported "facts of consciousness" for its evidence—though other, more *schwärmerisch* philosophies based upon allegedly superior powers of intuition are also possible. To base philosophy upon any appeal to immediately given "facts of experience"—whether mundane or extramundane—betrays a complete failure to understand that it is precisely the task of philosophy to "account for" or to "deduce" the system of these very facts. (This point was forcibly made by Fichte in his 1796 "Vergleichung des vom Herrn Prof. Schmid aufgestelleten Systems mit der Wissenschaftslehre."[53]) This criticism applies equally to appeals to inner and to outer experience, and thus it is a fundamental error to interpret the *Wissenschaftslehre* as a species of "introspectionism." To be sure, Fichte often appeals to the evidence of intuition, but, as we have seen, the objects intuited in the *Wissenschaftslehre* are never simply *given* to the philosopher, but must be deliberately *produced* through an act of imaginatively charged philosophical reflection. This difference is precisely what distinguishes *philosophy* from *psychology* and is why the *Wissenschaftslehre* is not guilty of "psychologism." (This distinction was first made explicit by Fichte in 1800 in his declaration "An das philosophische Publikum" and then further elaborated, a year later, in his "Antwortschreiben an Herrn Professor Reinhold."[54])

On the other hand, a simple call to base philosophy upon products of the creative imagination can all too easily open the door to the most extravagant and arbitrary fictions—as the history of philosophical Romanticism so richly attests. What separates the *Wissenschaftslehre* from something like Novalis's "magical idealism" is precisely Fichte's acute awareness of such a danger and his attempt to avert it by insisting that the productive imagination can play a valid role in philosophical reflection *only in constant conjunction with inference and intuition*. Fichte certainly believed that imagination is required in philosophy, but without the iron discipline of analytic thinking and the "reality check" provided by careful observation, a mere philosophy of imagination would be an even greater monstrosity than a philosophy based upon "thinking qua thinking" or upon a direct appeal to "the facts of consciousness."

What is most characteristic of the method of the early *Wissenschaftslehre*, therefore, is precisely the way in which it *balances* the equally important contributions of thinking, intuition, and imagination and attempts to *check* the excesses of each by the presence of the others. The "mixed method" of the *Wissenschaftslehre* might thus also be described as "a method of checks and balances." All three of these elements are combined in the concept of *philosophical reflection*, and thus all three are required for a philosophical justification—that is, a *transcendental deduction*—of ordinary consciousness. A philosophical "deduction" is an attempt to explain and and to justify certain everyday claims concerning, e.g., the objectivity of the world or of the moral law, by revealing the a priori conditions of the same (which, for a philosopher such as Fichte, turns out to mean: by relating these same features of experience to certain structures and activities of the I). Though logical inference and pure analysis are always involved in such a process of justification, a philosophical or transcendental deduction is by no means the same sort of purely formal "deduction" that one encounters within logic.[55] In addition, a philosophical deduction also has to rely upon reflectively purified self-observation, as well as upon a certain amount of creative guesswork. As such, the method of philosophy is an *ars inveniendi*,[56] and doing philosophy has more in common with solving a mystery than it does with calculating a sum.

What, however, is supposed to guarantee the *objective necessity*, that is to say, the *truth* of such peculiar "deductions"? Though a full answer to this question would require an examination both of Fichte's ultimately *practical* conception of truth and conviction and of his frank acknowledgement of the inescapable *circularity* of all transcendental explanations,[57] I will here confine myself to those factors related directly to the method just described. As in the case of Husserl's phenomenological "descriptions," what is supposed to guarantee the objectivity and universality of Fichte's "pragmatic history of the human mind" is that the latter must always be preceded by rigorous abstraction, in the course of which everything contingent and arbitrary is "thought away" from consciousness, along with the ordinary empirical contents of the same. In Fichte words, "the more the determinate individual is able to think himself way [*sich wegdenken kann*], the closer his empirical self-consciousness comes to pure self-consciousness."[58] Implicit in this assertion is the assumption that whatever can withstand such a reflective purge can be considered "universal" and "objective" in the sense that it is inseparable from consciousness as such and is hence "necessary." This is why Fichte can claim that the genetic acts that are reflectively posited in the *Grundlage* and the *Grundriß* are not mental acts of any particular individual person, but are the very acts through which self-consciousness originally posits itself as such.

A critic, of course, can always complain that such claims are groundless and that the so-called transcendental deduction of the universal character of subjectivity as such is no more than a "transcendental pretense," a contingent description of a particular empirical subject or of a particular kind of historical self—modern, Western, and male, for example.[59] The proper response to such criticism is the same as Fichte's response to his contemporay critics. The only way that anyone can become convinced of the truth of the propositions of the *Wissenschaftslehre* is to *posit them for oneself*, and the only way to evaluate the alleged results of Fichte's philosophical reflections is to *perform these acts of reflection and see what happens*. Unless a reader or critic is willing to do—or at least to attempt to do—just this, then there will be literally *nothing* for him to "see" in Fichte's writings.

In the end, therefore, Fichte's systematic treatises are no more than *provocations* designed to intice or to incite his readers and critics to think, to intuit, and to imagine for themselves. "The *Wissenschaftslehre* is not something that exists independently of us and without our help. On the contrary, it is something that can be produced only by the freedom of our mind, turned in a particular direction."[60] These words written in 1794 have lost none of their force, for it is just as true now as it was then that "[p]hilosophy is not something that floats in our memory or is printed in books for us to read; instead, it is what has stirred and transformed our spirit and has ushered it into a higher, spiritual order of things. It is something that has to exist within us."[61]

NOTES

1. From an unpublished fragment from 1800, in *J. G. Fichte-Gesamtausgabe der Bayerischen Akademie der Wissenschaften*, eds. Reinhard Lauth, Hans Gliwitzky, and Erich Fuchs (Stuttgart–Bad Cannstatt: Frommann-Holzboog, 1964ff.), II/5: 438.

2. See Daniel Breazeale, "Philosophy and the Divided Self: On the the Exis-

tential and Scientific Tasks of the Jena *Wissenschaftslehre*," *Fichte-Studien* 6 (1994): 117–47, and "The 'Standpoint of Life' and 'The Standpoint of Philosophy' in the Jena *Wissenschaftslehre*," in *Transcendentalphilosophie als System: Die Auseinandersetzung zwischen 1794 und 1806*, ed. Albert Mues (Hamburg: Felix Meiner, 1989), pp. 81–104.

3. Letter to Reinhold, July 4, 1797 (Fichte, *Gesamtausgabe*, III/3: 69).

4. See Max Wundt, *Fichte-Forschungen* (Stuttgart: Frommann, 1929), pp. 3–141; and Luigi Pareyson, *Fichte: Il sistema della libertà*, 2d ed. (Milan: Mursia, 1976), pp. 249–53.

5. J. G. Fichte, *Ueber den Begiff der Wissenschaftslehre* (1794), in *Gesamtausgabe*, I/2: 112; *Johann Gottlieb Fichtes Sämmtliche Werke*, ed. I. H. Fichte (Berlin: Viet & Co.,1845–46), I p. 38.

6. J. G. Fichte, *Ueber den Begriff der Wissenschaftslehre*, I/2, 115; *Sämmtliche Werke*, I, 42.

7. *Kritik der reinen Vernunft*, A712/B740–A758/B706.

8. Fichte, *Ueber den Begriff der Wissenschaftslehre*, 113–15; *Sämmtliche Werke*, I, 39–41.

9. Fichte, *Ueber den Begriff der Wissenschaftslehre*, 140–42; *Sämmtliche Werke*, I, 70–72. See too "Ueber den Unterschied des Geistes und des Buchstabens in der Philosophie," in *Gesamtausgabe*, II/3: 324–25 and *Grundlage der Naturrechts*, in *Gesamtausgabe*, I/3: 317 and 316n; *Sämmtliche Werke*, III, 5–6 and 5–6n.

10. On the distinction between an empty *Formular-Philosophie* and a "eine reele philosophische Wissenschaft," see Fichte, *Naturrecht*, 313–18.

11. J. G. Fichte, "Rezension Aenesidemus," in *Gesamtausgabe*, I/2: 46; *Sämmtliche Werke*, I, 8.

12. Fichte, *Ueber den Begriff der Wissenschaftslehre*, 148–49; *Sämmtliche Werke*, I, 79–80.

13. Fichte, "Ueber den Unterschied, des Geistes und des Buchstabens in der Philosophie," 330.

14. See J. G. Fichte, *Grundlage der gesammten Wissenschaftslehre*, (1794–95), in *Gesamtausgabe*, I/2: 404, 408, 432; *Sämmtliche Werke*, I, 271, 276, 305.

15. Fichte, *Ueber den Begriff der Wissenschaftslehre*, 47; *Sämmtliche Werke*, I, 77.

16. See Ernst Platner, *Philosophische Aphorismen*, Erster Teil (Leipzig: im Schwickertschen Verlag, 1793); Erstes Buch, " Pragmatische Geshichte des menschlichten Erkenntnißvermögens, oder Logik" [rpt. in Fichte, *Gesamtausgabe*, I/4S]. For Fichte's comments on the meaning of the term "pragmatische Geschichte" in this context, see *Gesamtausgabe*, II/4: 46, 52. Regarding the description of philosophy as the "the systemtatic history of the human mind in its univesal modes of acting," see "Ueber den Unterschied, des Geistes und des Buchstabens in der Philosophie," 334.

17. Fichte, *Ueber den Begriff der Wissenschaftslehre*, 142; *Sämmtliche Werke*, I, 72. See too *Gesamtausgabe*, I/2: 146; *Sämmtliche Werke*, I, 77.

18. See § 4 of the introduction to the *Kritik der Urteilskraft*.

19. See, e.g., Fichte, "Ueber den Unterschied, des Geistes und des Buchstabens in der Philosophie," 335–36.

20. "No abstraction is possible without reflection, and no reflection is possible without abstraction" (Fichte, *Ueber den Begriff der Wissenschaftslehre*, 138; *Sämmtliche Werke*, I, 67–68). See too Fichte, *Grundlage der gesammten Wissenschaftslehre*, 142; *Sämmtliche Werke*, I, 72.

21. See Fichte, *Ueber den Begriff der Wissenschaftslehre*, 142–48; *Sämmtliche Werke*, I, 72–79.

22. See especially Fichte, "Ueber den Unterschied, des Geistes und des Buchstabens in der Philosophie," 324–26.

23. Fichte, "Ueber den Unterschied, des Geistes und des Buchstabens in der Philosophie," 330.

24. See Fichte, *Ueber den Begriff der Wissenschaftslehre*, 142; *Sämmtliche Werke*, I, 72.

25. See Fichte, *Ueber den Begriff der Wissenschaftslehre*, 142–43; *Sämmtliche Werke*, I, 72–73.

26. See Fichte, *Ueber den Begriff der Wissenschaftslehre*, 146; *Sämmtliche Werke*, I, 77, as well as the 1795 essay "Ueber Belebung und Erhöhung des reinen Interesse für Wahrheit" (*Gesamtausgabe*, I/3: 83–90; *Sämmtliche Werke*, VIII, 342–52).

27. Fichte, *Ueber den Begriff der Wissenschaftslehre*, 110, 126, 117; *Sämmtliche Werke*, I, 30, 54, 44.

28. See, e.g., "Vergleichung des vom Herrn Prof. Schmid aufgestelleten Systems mit der Wissenschaftslehre" (Fichte, *Gesamtausgabe*, I/3: 254; *Sämmtliche Werke*, II, 442–43).

29. See Fichte, "Ueber den Unterschied, des Geistes und des Buchstabens in der Philosophie," 324. "Spirit as such is the same thing as 'the power of productive imagination' " ("Ueber den Unterschied, des Geistes und des Buchstabens in der Philosophie," 316).

30. Fichte, "Ueber den Unterschied, des Geistes und des Buchstabens in der Philosophie," 328.

31. Fichte, "Ueber den Unterschied, des Geistes und des Buchstabens in der Philosophie," 325.

32. Fichte, *Naturrecht*, 316; *Sämmtliche Werke*, III, p. 5; and "Ueber den Unterschied, des Geistes und des Buchstabens in der Philosophie," 327.

33. Letter to K. L. Reinhold, July 2, 1795,

34. J. G. Fichte, "Die von Fichte gehaltene Schulßvorlesung," in *Gesamtausgabe*, I/4: 413.

35. Fichte, *Grundlage der gesammten Wissenschaftslehre*, 255; *Sämmtliche Werke*, I, p. 91.

36. J. G. Fichte, *Grundriß der Ergenthümlichen der Wissenschaftslehre in Rücksicht auf das theoretischen Vermögen*, in *Gesamtausgabe*, I/3, 143; *Sämmtliche Werke*, I, 331. Regarding the "experimental" character of this portion of the *Grundlage*, see, e.g., *Grundlage der gesammten Wissenschaftslehre*, 262, 269–70; *Sämmtliche Werke*, I, 100,

108–109. This full "dialectical/deductive" strategy of "seeking out" the highest possible synthesis is actually carried out only in part 2 of *Grundlage der gesammten Wissenschaftslehre*, whereas in part 3 Fichte jumps directly to the "major antithesis" of the practical part. See *Grundlage der gesammten Wissenschaftslehre*, 386; *Sämmtliche Werke*, I, 247.

37. Fichte, *Grundlage der gesammten Wissenschaftslehre*, 316; *Sämmtliche Werke*, I, 152.

38. For examples of Fichte's frequent use in *Grundlage der gesammten Wissenschaftslehre* of the term *aufsuchen* or "seeking out" to describe an essential part of the method of philosophy, see *Gesamtausgabe*, I/2: 255, 275, 283–84; *Sämmtliche Werke*, I, 91, 114, 123–24.

39. On the role of inference in this process, see Fichte, *Grundlage der gesammten Wissenschaftslehre*, 269, 279, 316, 342; *Sämmtliche Werke*, I, 106, 107, 119, 163, 194.

40. Fichte, *Grundlage der gesammten Wissenschaftslehre*, 262; *Sämmtliche Werke*, p. 100. A corresponding power—and hence a corresponding "original fact of the mind"—is also deduced in part 3: namely, the I's original striving to determine some possible object.

41. See Fichte, *Grundlage der gesammten Wissenschaftslehre*, 364; *Sämmtliche Werke*, I, 121.

42. "We place the previously derived fact at the foundation and then we see how the I might be able to posit within itself this same fact. The latter positing is also a fact, and it must also by posited by the I within itself. We continue in this manner until we have arrived at the highest theoretical fact—that is, at that fact by means of which the I consciously posits itself as determined by the Not-I" (Fichte, *Grundriß*, *Gesamtausgabe*, I/3: 145; *Sämmtliche Werke*, I, 333). See too Fichte, *Grundriß*, *Gesamtausgabe*, I/2: 424; *Sämmtliche Werke*, I, 295 and Fichte, *Grundriß*, *Gesamtausgabe*, I/3: 170; *Sämmtliche Werke*, I, 363. As Fichte puts it, the method of the presentation here shifts from that of indirect, *apogogic* inferential/dialectical demonstration to that of direct demonstration by means of genetic description. See Fichte, *Grundriß*, *Gesamtausgabe*, I/2: 405 and 408; *Sämmtliche Werke*, I, 271 and 276.

43. Fichte, *Grundlage der gesammten Wissenschaftslehre*, 253; *Sämmtliche Werke*, I, 89.

44. See, e.g., Fichte, *Grundlage der gesammten Wissenschaftslehre*, 404; *Sämmtliche Werke*, I, 271.

45. Fichte, *Grundlage der gesammten Wissenschaftslehre*, 365; *Sämmtliche Werke*, I, 222.

46. Fichte, *Grundlage der gesammten Wissenschaftslehre*, 409; *Sämmtliche Werke*, I, 277. See too *Ueber den Begriff der Wissenschaftslehre*, 141; *Sämmtliche Werke*, I, 71.

47. Fichte, *Grundriß*, 146; *Sämmtliche Werke*, I, 335.

48. Fichte, *Grundlage der gesammten Wissenschaftslehre*, 362–63; *Sämmtliche Werke*, I, pp. 219–20.

49. Fichte, *Grundlage der gesammten Wissenschaftslehre*, 363; *Sämmtliche Werke*, I, 221.

50. See Fichte, *Grundlage der gesammten Wissenschaftslehre*, 420; *Sämmtliche Werke*, I, 290–91, where Fichte describes how the philosopher tries to retrace freely—in the person of the I he has constructed for himself and whose acts he now "observes"—the same path he posits himself as having previously traveled "out of necessity."

51. Fichte, *Grundlage der gesammten Wissenschaftslehre*, 415; *Sämmtliche Werke*, I, p. 284.

52. Fichte, *Gesamtausgabe*, I/6: 433–450; *Sämmtliche Werke*, II, 490–503.

53. Fichte, *Gesamtausgabe*, I/3: 235–66; *Sämmtliche Werke*, II, 421–58.

54. See Fichte, *Gesamtausgabe*, I/6: 458, and I/7: 296; *Sämmtliche Werke*, II, 509–10.

55. This explains why contemporary attempts to explicate the purely logical form of "transcendental arguments" have failed so spectacularly to capture the true character of transcendental philosophy.

56. For the use of this term to describe Fichte's philosophical method, see Edith Düsing, *Intersubjektivität und Selbstbewußtsein* (Köln: Dinter, 1986), pp. 255–56.

57. See Daniel Breazeale, "Certainty, Universal Validity, and Conviction: The Methodological Primacy of Practical Reason within the Jena *Wissenschaftslehre*," in *New Perspectives on Fichte*, eds. Tom Rockmore and Daniel Breazeale (Amherst, N.Y.: Humanity Books, 1995).

58. Fichte, *Grundlage der gesammten Wissenschaftslehre*, 383; *Sämmtliche Werke*, I, 244.

59. For an example of such a criticism, see Robert C. Solomon's *Continental Philosophy since 1750: The Rise and Fall of the Self* (Oxford: Oxford University Press, 1988).

60. Fichte, *Ueber den Begriff der Wissenschaftslehre*, 119; *Sämmtliche Werke*, I, p. 46.

61. Fichte, "Ueber den Unterschied, des Geistes und des Buchstabens in der Philosophie," 332–33

DEDUCTION
IN THE
GRUNDLAGE

FICHTE'S DEDUCTION OF REPRESENTATION IN THE 1794–95 *GRUNDLAGE*

Steven Hoeltzel

INTRODUCTION

One important challenge inextricably bound up with Johann Gottlieb Fichte's attempt to disclose the articulate-but-unitary, active-cum-cognitive originary structure of the transcendental subject concerns the possibility of constructing "a pragmatic history of the human mind":[1] an account of the operations of this structured complex of proto-discursive activities that intelligibly derives from their manifold workings the actual constitution and character of an empirical subject's ordinary discursive, representational states of awareness. Absent such an account, after all, it is not clear that the *Wissenschaftslehre* can make good upon the rather startling explanatory pretensions broadcast in Fichte's proclamation that "everything which occurs in our mind can be completely explained and comprehended on the basis of the mind itself."[2] Such a thoroughly idealistic explanation of conscious experience is lacking unless a "genetic deduction of what we find in our consciousness"[3] is successfully carried out.

However, there may be good reason to suspect that such a deduction must be an impossibility, given the very nature of the Fichtean enterprise. This project, after all, involves reliance upon regressive arguments which purport to isolate a cluster of forbiddingly abstract, preconscious activities that in some way lay the foundation for our everyday empirical states of mind. But the activities thereby singled out are so utterly different from anything of which we are ordinarily cognizant that the emergence of full-fledged empirical consciousness from such meager beginnings proves extremely difficult fully to understand. That is to say, even if Fichte's regressive arguments go through, in themselves they provide us with no obvious way of understanding how our transcendental *explanantes* actually give rise to our experiential *explananda*. We unveil a cluster of abstract conditions that operate as necessary ingredients in the constitution of representational awareness, but we do not *thereby* attain any understanding of the actual cooperative coordination of these conditions in this constitution process. Unless some way of spanning this apparent gap between the transcendental and the empirical is devised, there is, it seems, no question of providing a genuine "pragmatic history" of world-related mental states.

The "Deduction of Representation" in the 1794–95 *Grundlage* is clearly a provisional attempt to make good upon this requirement.[4] Hence the actual details of this "deduction," bewildering though they may be, are of considerable systematic importance. For it is crucial to Fichte's overall project that this explanatory gap be filled in, as I demonstrate in more detail below (I). Once the nature of the problem and its significance for the *Wissenschaftslehre* as a whole have been rendered more precise, I closely survey the intricacies of the deduction itself, first reading the myriad details of the argument in the light of its systematic function (II), then briefly— and somewhat critically—assessing the argument's success (III).

I. The Deduction of Representation and the Structure of World-Related Selfhood

Following the lead of Karl Leonhard Reinhold's *Elementarphilosophie*, Fichte proposes to construct a transcendental theory of human subjectivity in which every element admits of systematic derivation from a single, self-certifying first principle (*Grundsatz*).[5] The *Wissenschaftslehre*, however, begins with and evolves out of Fichte's conviction that Reinhold's purportedly foundational "Principle of Consciousness" (*Satz des Bewußtseins*) cannot function as philosophy's true *Grundsatz*, since it fails to disclose experience's ultimate nonempirical ground and therefore leaves the elementary structure of world-relatedness radically unexplained. Instead, Reinhold's basic proposition merely *describes* the articulate configuration of other-directed awareness in general, the basic form that all representational consciousness has *for* the subject doing the representing: "In consciousness, the subject distinguishes the representation from both the subject and the object and relates it to them both."[6]

This principle functions primarily not as an ontological claim, but as a technical specification of the most basic way in which experience is ordered by and for the conscious subject. Notice that the proposition figures world-related awareness as internally structured in a triadic but essentially bipolar way: the experiential field within which representation occurs is categorially circumscribed by subject and object; a particular mental content mediates these two extremes by virtue of being figured by the subject as related to both though identical to neither. More importantly, though, note that the Principle of Consciousness qua *Grundsatz* treats the presence of this articulated structure as a brute fact, not to be explained by reference to any other, more basic, fact or activity. Subject, object, and representation are taken to be simply present as such (albeit always in an empirically determinate way) for the conscious mind, and their presence as such is taken to be *the* most primitive ingredient of all determinate states of awareness.

Hence Fichte's denial of the principle's philosophical ultimacy. For even if we allow that "[t]he opposite of what this principle asserts is not even conceivable,"[7] Reinhold's proposition fails to explain how and why world-related selfhood must come into being structured in just this way. "If consciousness exists," Fichte maintains, "then this is itself a fact, and it must be derived like any other fact."[8] More precisely, transcendental philosophy must locate among the a priori acts of the subject the genesis of the differentiated structure that the Principle of Consciousness describes. This proposition "is a theorem which is based upon another first principle, from which, however, the Principle of Consciousness can be strictly derived, a priori and independently of all experience."[9] The mind's grasp of the basic, organizing moments of experience cannot be grounded in or explained by any world-related awareness, because these elements are among the ultimate preconditions of any such awareness. Hence the subject must have somehow produced these abstract fundaments and effected their myriad relations prior to any empirical discovery of their material specifications.

As Fichte rather cryptically puts it in the *Aenesidemus* review, "The subject and object do indeed have to be thought of as preceding representation, but not in consciousness qua an empirical mental state, which is all that Reinhold is speaking of. The absolute subject, the I, is not given by empirical intuition; it is, instead, posited by intellectual intuition. And the absolute object, the not-I, is that which is posited in opposition to the I."[10] That is to say, world-related awareness is ultimately grounded in preconscious activities that originate experience's organizing elements a priori. The transcendental subject—the I—originally constitutes itself via an unconditioned act

(*Tathandlung*) of self-positing.[11] Then, in a logically subsequent but empirically unconditioned act, the I originally sets itself up to encounter an object as distinct from itself by counterpositing a not-I to itself.[12] And in order that the posited I is not annihilated by the positing of the not-I as such, I and not-I are posited as mutually limiting one another.[13]

In its inference from the manifest facts of discursive subjectivity to the necessity of these preconscious acts of protodiscursive positing, the *Wissenschaftslehre* purports to uncover "the basis or foundation of all experience," which basis "must necessarily lie *outside of all experience*."[14] But the abstract cluster of originary activities thus disclosed comprises only the rudiment of a complete explanation of experience and its organizing structure. Fichte has yet to render intelligible the transition from this precognitive agency to outright world-related awareness, in which the subject does not merely posit, but also *represents*, and in which its protodiscursive commitment to an abstract I and not-I undergirds an empirical subject's further grasp, via a determinate representation, of some particular object. We know that some such enrichment of the I must occur, but until we understand precisely how it occurs, the promised a priori deduction of the principle of consciousness is lacking, as is the *Wissenschaftslehre*'s vaunted proof that "everything which occurs in our mind can be completely explained and comprehended on the basis of the mind itself."[15]

Moreover, such an explanation is not only required in order that Fichte's theory may live up to some of its more daring assurances. It is also necessary, owing to certain of Fichte's other commitments, as an indirect demonstration—arguably the only sort possible—of the correctness of the *Wissenschaftslehre*'s radically idealistic orientation. On Fichte's view, a philosophical account of experience must posit either the protodiscursive subject (the spontaneously self-positing I) or the extra-experiential object (the thing in itself), but not both, as the ultimate explanatory ground.[16] The explanation of experience must then proceed by way of rigorous systematic deduction from the proposed first principle: the spontaneously self-positing I in Fichte's theory, the efficacious action of the thing in the system of "dogmatism."[17] Hence at the level of philosophical theory, the *Wissenschaftslehre* simply begs the question against any attempt systematically to explain experience on a naturalistic basis—and, of course, vice versa.[18] Fichte seems to maintain, however, that the supremacy of idealism can be demonstrated at the *meta*-theoretical level, precisely insofar as it can be shown that idealism succeeds, while naturalism fails, at systematically deriving experience from its extra-experiential ground.[19]

It is therefore crucial to Fichte's overall project, which is inaugurated by the contention that Reinhold's Principle of Consciousness demands an a priori derivation which only idealism can provide, that such a systematic deduction be furnished. And the Deduction of Representation looks to be the one place in the *Grundlage* where Fichte can claim to have supplied this need, since this exposition begins with the absolutely self-positing I (qua originally determined in an important way) and culminates in an empirically representing, world-related self.

II. THE DEDUCTION OF REPRESENTATION IN THE 1794–95 GRUNDLAGE

The overarching project of the theoretical division of the *Grundlage* is to provide the demanded derivation of world-related awareness out of the tendencies and tensions latent in the dynamic activity structure that is the transcendental subject as described by the *Wissenschaftslehre*'s first principles. The Deduction of Representation proper[20] encompasses both the culmination of this enterprise and an important shift in its methodology.[21] Prior to the inception of this "deduction," the argument of the theoretical section is a systematic, dialectical unfolding of the necessary conditions of the protodiscursive act whereby the I originally constitutes itself as an intelligence: the I's spontaneous positing of itself qua limited by the not-I.[22] Mediating determinations by virtue of which these absolute opposites may be posited together in a single act—i.e., categories—are required to explain *how* such a synthesis can take place. And "*a primordial fact occurring in our mind*"[23]—the notorious *Anstoß*—is necessary to explain *why* the I spontaneously counterposits a not-I in the first place.[24]

Once the relevant categories and the original determination of the precognitive, self-positing subject by the *Anstoß* have been deduced, Fichte maintains, "All the elements necessary to explain representation . . . have been grounded and established; and from now on we thus have to do no more than apply and connect what has been proved so far."[25] The dynamic self-constitution of world-related awareness on the basis of the protodiscursive rudiments thus far uncovered is "from now on the object of our philosophic reflection,"[26] which need do no more than "calmly follow the course of events."[27]

This movement from dialectical derivation to "philosophic reflection" constitutes the aforementioned methodological shift. In any act of philosophical reflection, as Fichte uses this term, "Something which in itself is already form (i.e., the necessary action of the intellect) is incorporated as content into a new form (the form of knowledge or consciousness)."[28] Hence in "deducing" representation as such, the philosopher begins by intending representations of the protodiscursive activities of the I. These are one and all representations of a spontaneous dynamism, but the acts of the mind invariably occur in determinate, law-governed ways, and therefore "present a system for any observer."[29] Thus the reflecting philosopher should find that one representation—one represented act of the mind—*necessarily* gives way to a certain new representation, and no other, until the highest level of determinacy is reached: reflection grasps a representation *of* representational consciousness, structured as Reinhold's principle describes it.[30]

A number of important points follow from all this. First: although this quasi-phenomenological procedure does not proceed by way of formal inference, it still merits description as a "deduction." Fichte's operation derives, via a series of nonarbitrary steps, a result that is not initially manifest but which is nonetheless implicit in the beginning. Second: the philosopher's observation of the I's evolution will track this development within a conceptual frame of reference, organized by a basic subject-object bipolarity, which is not initially available to the evolving I.[31] Finally, because the I's modes of activity can actually enter consciousness "only insofar as and in the manner that they are represented,"[32] the philosopher's attempt to come to grips with the unfolding activity structure of transcendental subjectivity necessarily depends upon all manner of discursive contrivances: spatial, temporal, and quasi-mechanical representations, in particular.

This is nowhere more clear than in the passage with which Fichte opens the Deduction of Representation proper: "The endlessly outreaching activity of the I, in which nothing can be distinguished, precisely because it reaches into infinity, is subjected to a check; and its activity, though by no means to be extinguished thereby, is reflected, driven inwards; it takes exactly the reverse direction."[33] Here we have a spatially and quasi-mechanically cast representation of the unfolding of the I's articulated structure as it is first set into motion by the *Anstoß*, which impels the original, "infinitely outreaching" activity of the I to recoil back in the "direction" of its source.

What follows is a devout attempt to track and elucidate the actual details of Fichte's dense and tangled account of the further development of the I as this manifests itself to the philosopher's reflective gaze. This requires that I (the author, not the abstract agency under scrutiny) closely follow the text throughout, but I endeavor to maintain enough distance to introduce some much-needed clarity into the proceedings. By way of offering some additional orientation, I have also broken up the presentation into numbered sections with headings indicating the principal issues of concern.

A. Activity, passivity, and intuition

The *Anstoß* that "checks" the I's original activity does not cancel this activity entirely. For as Fichte points out, if the outgoing activity—the I's spontaneous self-positing—did not remain in some sense persistent as such, then the I would simply cease to be an I.[34] The activity which the *Anstoß* causes to be reflected back toward the I must, therefore, retain some degree of outward orientation at the same time: it must be an activity whose structure contains a kind of duality or bidirectionality. As Fichte describes it, there thus arises "a twofold direction of the I's activity, at variance with itself"—a passivity (the recoiling activity), and an activity (the outgoing activity), both of which are "one and the same state of the I."[35] This single state in which opposing tendencies are seamlessly intermingled is, Fichte maintains, an activity of *imagination*.[36] More precisely, it is a state of *intuition*: an activity that depends upon a passivity (the reflection of the I's original activity occasioned by the *Anstoß*), and a passivity that depends upon an activity (the original self-positing of the I).[37] Although the I's original activity of self-positing continually extends "beyond" the *Anstoß*, all intuitive activity occurs in the region "between" the *Anstoß* and the I, since beyond the *Anstoß* the I's original activity no longer meets with any resistance.[38]

Fichte is careful to interject at this point, however, that although the philosophical "observer" discovers the evolving I to be in a state of intuition, this is in no sense manifest to the I under observation. Intuition, he says, "is now determined, though only as such, for philosophical reflection; it remains completely indeterminate in regard to the subject, as an accident of the I; for in that case it would have to be distinguishable from other determinations of the I, which is not possible as yet. It is equally indeterminate in regard to the object, for in that case an intuited something would have to be distinguishable as such from something not intuited, which is likewise impossible so far."[39] At present, then, we have described only a very odd and unstable activity which oscillates between and surges toward each of two (as yet) indeterminate poles. Though we, the philosophical observers, recognize a subjective and an objective as poles between which the I's activity restlessly wavers, the observed I operates within no such frame of reference—it just *is* the very abstract activity structure under discussion. We have yet to see, that is, precisely how the intuiting I comes to be distinguished from the intuited not-I *for* the I under observation. The abstract activity structure so far examined will therefore have to somehow yield further articulation and determination.

That the observed I intuits an intuited object *for itself* requires, quite simply, that the I *posit* itself as intuiting.[40] This requires above all, Fichte argues, that "in intuition it posits itself as *active*."[41] The I, in other words, must come to identify itself with only the active aspect of the bidirectional activity of intuition. This means, of course, that the passive dimension of this activity must be attributed to some other source. That is, the I must posit "something opposite to itself that is not active therein, but passive."[42]

B. The initially unreflected production of the intuited

In what manner, then, is an intuited item opposed to the I? Necessarily as a not-I.[43] And because the act of the I that posits the intuited not-I is not a *self*-directed activity, this act is "*not a reflection*, not an inward-directed but an outward-directed activity, and thus, so far as we can see at present, a production. The intuited, as such, is produced," via the productive imagination.[44] But because *this* activity is not reflected or attributed to the observed I except by the observer, the observed I is not conscious of the activity whereby it produces the intuited as such.[45]

The act whereby the I posits itself as intuiting therefore must involve a fair degree of internal complexity. It must involve both (1) "an activity of intuiting which the I ascribes to itself by means of reflection," and (2) "another [activity—one productive of the intuited qua intuited] which it does not so ascribe."[46] But though we now see what is required of the I that would posit itself as intuiting, we have yet to see precisely how the observed I could actually perform this act. For this act requires that the I be capable of *self*-ascription and *other*-directedness, and we have yet to see how these capacities necessarily unfold out of the I's originary activity structure. At this point, the observed I just *is* a mere buzzing tangle of directionally conflicting acts; it has no fixed conceptual landmarks (subjective and/or objective) with reference to which the meanings of its various activities could become intelligible *to it*.

C. The need to reflect the intuiting activity

The problem, spelled out in a bit more detail, is this: the *Anstoß* has occasioned the reflection of the original, "infinitely outgoing" activity of the I back in the direction of its source. This produces a state of intuition in this I—but by itself this does not suffice to give rise to intuition properly so called, since the resulting state is not yet a state of intuition *for* this I. In order to intuit, the I must posit itself as intuiting. And in order to do this, the I must spontaneously generate *another* self-directed activity, an act of reflection distinct from the activity already recoiling back toward the I. "If these twin tendencies are not distinct," Fichte argues, then "no intuition at all is reflected, there being a mere repetition of the intuiting in exactly the same manner."[47]

In order to account for the I's positing itself as intuiting, then, we need to show how and why the second spontaneous, reflective act should result in something *more for* the I (namely, the reflection of the I's *intuiting* back upon the I) than does the first activity which recoils upon the I (and gives rise to intuition, but not intuition *for* the I). But at this point, the I lacks the resources even to tell the two acts apart: for this I, according to Fichte, "it is one and the same activity of the I . . . [insofar as] it has the same direction."[48] Of course, at the level of philosophical reflection it is clear that "the first is reflected by a mere check from without, while the second occurs through absolute spontaneity."[49] The question, though, is what originally grounds our ability to make this reflective discrimination—"how the human mind originally comes to make this distinction between a reflection of activity from without, and another reflection from within."[50] The problem, then, is to discover how the I comes to conceptually fix its initially indeterminate activities in such a way that "without" and "within" can have an unequivocal meaning for it.

D. Understanding and the stabilization of intuition

According to Fichte, a clearer determination of I and not-I as subject and object of intuition requires first of all that intuition be "stabilized"—otherwise, "we clearly have no fixed point, and are revolving endlessly in a circle."[51] To stabilize intuition, we are told, means to fix it or determine it in such a way that not just one aspect of the intuiting activity (intuitant or intuited), but indeed this activity generally, can be reflected in the direction of the I—so that the I can figure itself *as* the intuitant of something intuited. However, intuition is by its very nature unstable: it consists essentially in "a wavering of the imagination between conflicting directions."[52] The requirement that intuition be stabilized would therefore seem to amount to the demand that it be canceled out entirely. But because this clearly will not do, it must be the case that intuition can at least be divested of all but traces of the opposed directions—"consisting of neither but containing something of both."[53]

As Fichte describes it, *understanding* does the job of stabilization, arresting the wavering of the imagination via the apprehension and conception of its intuitive products.[54] This operation of understanding, therefore, is yet another necessary condition of the I's positing itself as intuiting. For as we have seen, this act of positing can take place only in so far as the intuitive activity as a whole can be reflected back toward the I, which requires in turn that the intuition be stabilized by understanding's exertion of its calming power. Fichte rather cryptically makes this point as follows: "In understanding alone (albeit first through the power of imagination) does reality *exist*; it is the faculty of the *actual*; the ideal first becomes real therein. . . . Imagination produces reality, but there *is* no reality therein; only through apprehension and conception in the understanding does its product become something real."[55]

E. The unreflected counterposition of the not-I.

After stressing the crucial role of understanding, Fichte recurs to the topic of the intuitive activity which must be reflected back in the direction of the I in order that the I may posit itself as intuiting. Insofar as this activity is reflected, he maintains, it is reflected as extending outward only to the point at which it abuts the *Anstoß*, at which point it is "limited and determined."[56] The unconditioned act of reflection thus described, however, is not itself reflected by the observed I; it is reflected upon only by the philosophical observer. So while this activity brings it about that limitation and determination are *for* the observed I, it is not *itself* manifest to this I. For this reason, the I has to blame its limitation and determination on something besides its own activity—so that, as Fichte explains it, "The limitation . . . is posited counter to the I and attributed to the not-I. Into the infinite beyond there is projected a determinate product of the absolutely productive imagination, by means of a dark, unreflected intuition that does not reach determinate consciousness; and this product sets limits to the power of reflected intuition. . . ."[57]

It is, moreover, via this "unreflected" counterpositing of the not-I, that the I is first determined *as* an I—"which first makes possible the logical subject of the proposition: the I intuits."[58] Absent the counterpositing of the not-I, the observed I fills all logical space for itself; there is no not-I with reference to which the I can figure itself *as* an I in distinction from something else. As we will see, though, this I, held fast at the level of its original, unreflected positing of the not-I, is still not yet an I that is able to posit itself as intuiting. To be sure, the counterpositing of the not-I furnishes some ground upon which the observed I might discern some difference between itself qua intuitant and the not-I qua intuited. But this I will find, initially, that these items are so conceptually interdetermined that any distinction between them tends to collapse, preventing the I from seizing upon (positing) just one of them as identical to itself.[59]

F. The continuing search for a ground of distinction

As we have already noted, insofar as the observed I is to posit itself as an intuitant, it must posit itself as active in intuiting. And in order to do this, Fichte now adds, this I must be able to distinguish its intuiting activity from the various other activities with which this one is intermingled.[60] As yet, for the observed I, the only salient feature of an activity is its direction; and the activity which is appropriately opposed to the I's activity of intuiting is that recoiling activity which tends from the *Anstoß* toward the I, "occasioned by reflection from without, and conserved in the understanding."[61] But this activity, qua activity, foils the I's attempt to posit only *itself* as active in intuition.

Nor can the I recognize the activity which opposes its intuiting activity as the activity of that which is not-I, because the not-I is never actually intuited. Rather, "A thing of this kind lies out beyond [the *Anstoß*] as an absolute product of the I's activity."[62] That which can be apprehended as real (the intuited qua stabilized in understanding) must inhabit the region between the I and the point at which the I's activity is originally "checked." The unintuited not-I, therefore, cannot be present to the intuiting I as a resource for distinguishing intuitant from intuited.

Still, the I requires some way of effecting a distinction between itself and the not-I. "Both must be determined, for the I is to posit itself as the intuitant, and to that extent oppose itself to the not-I."[63] As yet, though, no workable ground of distinction is to be found: intuited and intuitant are interdetermined in terms of activity and passivity, reality and negation, and simultaneous positing. Determine or posit one, and the other is thereby determined or posited as well.[64] The observed I cannot (yet) do what it must: get one or the other of these opposites solidly in its grasp. Thinking either one (in general or under some determination) requires thinking the other (in general or under some determination), and so on *ad nauseam*. This I's attempt to rein in its activities between a stable pair of fixed objective and subjective elements is, as yet, unsuccessful. But for this very reason, Fichte says, "one of the two must be determined *by itself*, and not by the other, since otherwise there is no exit from the circle of interdetermination."[65]

G. The activity-structure of intuition

Perhaps, he then suggests, there is some structural feature inherent in intuition itself that will provide the observed I with the resources to posit itself as active in intuiting without being drawn into this circle. The intuitant qua active, he argues, is engaged in an *objective* activity: "an activity to which there corresponds a passivity in its opponent."[66] Since this activity, as we have seen, depends upon the I's primordial act of self-positing being originally determined by an *Anstoß*, this objective activity is always conditioned by *pure* activity, "activity absolute and in general."[67] Intuition, as an activity *of the intuitant*, is therefore an activity that is possible only under a certain condition: namely, that the "pure" activity of the absolute I be originally determined by an *Anstoß* that gives rise to a(n initially unreflected) state of intuition in the I. But that there is an intuited something *for* this is I clearly depends, ultimately, upon the very same condition: where this condition is not met, there is no intuition at all, and hence no distinction between intuitant and intuited *for* the I. There is, instead, only the unconditioned act of infinite self-assertion, the *Tathandlung*.[68]

That intuition occurs only under a certain circumstance is therefore of no help in effecting the required discrimination, since the condition in question attaches equally to both intuitant and intuited. But perhaps, Fichte suggests, the more specific conditions under which intuition occurs are sufficiently different with respect to each of these opposites to provide the observed I with the ground of distinction that it needs.[69] His exploration of this option begins with an instructive account of the conditions under which the I originally comes to intuit.

Two very different types of activity, Fichte contends, are united in the I qua intuitant—though, to be sure, none of this is originally manifest to the observed I itself. First, there is the I's uncoerced "self-affection": the infinitely outgoing activity of the I is originally determined by the *Anstoß*, but its subsequent unfolding of further articulations is not necessitated thereby. Nevertheless, this unfolding would not occur in the absence of the *Anstoß*, so the I's free self-determination does depend upon a kind of compulsion. As Fichte puts it, "Spontaneity can reflect only under the condition of a reflection already brought about by an *Anstoß* from without; but even under this condition it is not *obliged* to reflect."[70] Put a bit differently, "Interaction between the self-affection of the intuitant and an affection from without, is the condition for an intuitant to be an intuitant."[71]

Things stand in precisely the same way, however, with the intuited object: its being intuited depends upon the same condition of reciprocal action. The intuited is passive in relation to the active spontaneity of the intuitant, but it is active vis-à-vis the latter's passive receptivity. Activity and passivity reciprocally determine both intuited and intuitant; hence, there is still no way in which the observed I can seize upon one without being driven to posit the other. It is still unclear, therefore, how this I can come to posit itself, uniquely, as active in intuition—which, remember, is required in order that the I may posit itself as intui*ting*, which in turn is necessary if the I is actually to intuit at all.[72]

Perhaps we can make some headway, Fichte suggests, if we can manage to "determine the share of one of the two components in the said interaction on its own account."[73] In particular, he suggests, it may be possible to fix precisely the intuitant's role in this interaction via careful scrutiny of its activity of self-determination—which seems unique among the activities currently at issue in that it involves "no corresponding passivity in the object," and thus seems to promise to help lead the observed I out of the circle of interdetermination.[74] This apparent promise, however, turns out to be empty. The self-reverting activity of the I determines the objective (intuiting) activity, insofar as intuition only occurs given the reflection back toward the I, occasioned by the *Anstoß*, of the I's original activity. The objective (intuiting) activity in turn determines the I's self-reverting activity, since it is precisely *in* intuiting that the I's activity takes on its self-reverting direction. Hence both modes of activity are interdetermined, and "[a]gain we have no fixed point of determination."[75] The observed I has no way of grasping *self*-determination as an activity which is discernible from its counterparts. For this reason, as far as this I can tell, "[t]he activity of the intuited in the interplay, so far as it relates to the intuitant, is similarly determined by a self-reverting activity, whereby it determines itself to operate upon the intuitant."[76] Once again, then, the observed I is afforded no way of telling intuited and intuitant apart.

H. Self-determination, other-determination, and thought

Fichte next proposes that we approach the I's activity of self-determination from a slightly different perspective. Insofar as the intuitant is self-determining, he argues, it performs "the determination, by reason, of an imaginative product fixed in the understanding: hence an act of *thought*."[77] The intuitant, in other words, determines itself to the *thinking* of an object, which object, qua thus determined, is an *object of thought*. Does the distinction between thinking and its object, which here seems to emerge, finally point the way clear of the circle of interdetermination? No. For according to Fichte, it follows from the nature of the intuitive interplay described just above that the object of thought, too, "is at once determined as *determining itself*, to an operation upon the intuitant."[78]

The observed I, therefore, still cannot distinguish the intuitant from the intuited. It is, at this stage of its evolution, caught up in the interdetermination of efficacy: the object of thought, figured as substrate of its own original and self-reverting activity, is construed as "the *cause* of a passivity in the intuitant, which is its *effect*."[79] Alternatively, as shown just above, the intuitant may be figured as the cause of a passivity in the object, insofar as it thinks this as a determinate object of thought. Hence, the observed I cannot uniquely single out intuited or intuitant by positing·one or the other as cause or effect: the relation of efficacy seems to run in both directions.

I. Judgment

Still, Fichte is not beaten yet. He now suggests that further determining the activity of thinking, "this activity of self-determination for purposes of determining a *determinate* object,"[80] will be of some help. In the intuitant, he argues, the activity of thinking an object qua determinate is itself "determined by an activity . . . of a kind that determines no object as a determinate (= A); an activity directed to no determinate object (and hence, in effect, to an object in general, as mere object)."[81] Such an activity is *free* vis-à-vis the object's determinacy: it may either reflect on the object, or abstract from it.[82] This activity, Fichte tells us, is the capacity of *judgment*, which may reflect upon or abstract from objects already posited in understanding.[83] The capacity for judgment, then, is an ability possessed by the observed I, even if this I is not yet explicitly aware of this fact. So perhaps the I's *exercise* of judgment will somehow allow it to make the distinction required in order, finally, to posit itself, uniquely, as active (in thinking, in this case).

As we have by now come to expect, however, things do not turn out so happily. For while determinative thinking is active in relation to the "mere object" as it exists for judgment, Fichte claims, the determinate object of thinking, "as object of thinking, and thus . . . as passive, is determined by something *not* thought, and thus by a mere thinkable (which must have the ground of its thinkability in itself, and not in what thinks, and thus is to that extent active, while what thinks is passive in relation thereto)."[84] The interdetermination of efficacy, in other words, is simply reiterated at this new level, with judgment's "thinker" and "thinkable" standing in for intuition's "intuited" and "intuitant." As Fichte concisely explains it, "What thinks is to determine itself to think of something as thinkable, and to that extent the thinkable will be passive; but then again the thinkable is to determine itself to be a thinkable; and to that extent what thinks will be passive."[85] The observed I still cannot get any one of these notions, which are perfectly distinct for philosophical reflection, solidly within its grasp. Therefore, Fichte resolves, we "must determine what judges to a further extent still."[86]

J. Pure reason

Fichte has already contended that in the observed I, the activity of thinking is itself determined by an activity "which has no object at all, an intrinsically nonobjective activity, opposed to the objective one."[87] He now argues that the latter activity, which grounds the possibility of abstracting from any given determinate object, itself depends upon the ability to abstract "from *all objects in general*."[88] And this ability, he urges, will somehow enable the observed I finally to make the distinction which has thus far eluded it, systematically frustrating its attempts to grasp itself *as* subject (intuitant, thinker) and its object *as* object (intuited, thinkable). "Such an absolute power of abstraction must exist," he contends, "if the required determination is to be possible; and the latter must be possible, if self-consciousness and consciousness of representation are to be possible."[89]

Fichte allows that there is no question of intuiting such a power, but argues that there is evidence for the existence of such a capacity in "the mere rule of reason in general, telling us to abstract . . . and hence this absolute power of abstraction is simply [pure] *reason* itself."[90] These claims are, of course, beyond the grasp of the observed I at this point. Nevertheless, this I possesses reason, and may therefore use it. And the exercise of this absolute power of abstraction quite literally *makes all the difference in the world for* this I. The subject-object polarities that our reflective survey has taken for granted throughout at last become manifest to this I as such. When everything objective has been abstracted away, Fichte argues, "We are left at least with what determines itself, and is determined by itself, the I or subject. . . . The I is now determined as that which remains over, after all objects have been eliminated by the absolute power of abstraction; and the not-I as that from which from which abstraction can be made by this same abstractive power: and thus we now have a firm point of distinction between object and subject."[91] The poles between which the observed I's various activities waver and surge are now, for the first time, rendered perfectly determinate *for* this I. These activities, the subject whose activities they are, and the objects of these activities, all acquire a determinacy for the I which they had lacked prior to this point, at which stage these items were one and all indistinguishable qua strictly interdetermined. No longer is the I doomed to be drawn into a circle of interdetermination in its attempt to figure the objects of its awareness as distinct from itself. Instead, "[e]verything that I abstract from, everything I can think away . . . is not my self, and I oppose it to my self merely by regarding it as something that I can think away."[92]

The I's ability to grasp itself (qua intuitant and thinker) as distinct from the objects of its awareness (the object qua intuited and/or thought) enables it to posit *itself*, uniquely, as the active principle in all intuiting and thinking. In other words, this capacity makes it possible for the I to posit itself *as* intuiting and thinking, and hence actually to intuit and actually to think—that is, to *represent, for itself.*

III. A CRITICAL REFLECTION

Does the Deduction of Representation live up to the demands that are placed on it? As I have already indicated, this is a question of momentous systematic importance for Fichte, given that the *Wissenschaftslehre* is born of his conviction that Reinhold's Principle of Consciousness requires an a priori derivation that only idealism can provide. It is also a question, however, which does not admit of a simple and unequivocal answer. Fichte pledges to derive the basic structure of world-related awareness from its preconscious rudiments "a priori and independently of all experience,"[93] but an apparent discrepancy between his actual procedure and certain claims that he makes on its behalf obscures the degree of independence of experience that he requires of a suitably rigorous deduction. In particular, it is not clear whether the bare fact that *there is* experience is allowed to exert any influence on the proceedings.

If the fact that experience occurs is not an admissible premise of the Deduction of Representation, then the attempted derivation can hardly be called a success. For in that case, the argument would have to demonstrate that every step in the progressive development of the I, from *Tathandlung* to world-related self, is strictly necessitated solely by the developments that have preceded it. And it seems clear that no such thing is demonstrated by this deduction—and hence within the *Grundlage* as a whole. Consider, for instance, the very first forward step that philosophical reflection observes. What reason are we given for thinking that the original reflection of the I's "infinitely outgoing" activity back in the direction of its source actually *necessitates* the simultaneous re-exertion of that outgoing activity? Fichte argues that if this re-exertion did not occur (if the I at some point ceased to posit itself), then the I would simply cease to be an I. Fair enough—but then why shouldn't the I cease to be an I at this very point? On what basis is it necessary that the original activity re-exert itself? (Of course, analogous questions, mutatis mutandis, are even more clearly applicable to later, more Byzantine stages of the subject's self-constitution.)

To fend off this difficulty, one can simply argue that any given step in the deduction, while not strictly necessitated by its transcendental forbears, is nonetheless a necessary *condition* of experience eventually arising for the I. But while this contention appears to comport nicely with Fichte's actual procedure, in so far as the various steps of the derivation seem to have no other real warrant, it is flatly contradicted by certain claims which Fichte makes regarding the *Wissenschaftslehre*'s deductions. First, in an essential supplementary text to the *Grundlage*, he makes the striking assertion that his system "is altogether unconcerned with experience and does not take it into account at all. The *Wissenschaftslehre* would have to be true even if there were no experience at all."[94] This would certainly seem to obviate any need to refer to experience in the construction of the theory. Moreover, he clearly regards the acts of the mind as occurring in determinate, law-governed ways,[95] such that a reflective observer of these acts is not called upon to make any contribution to their evolution by prodding them in the appropriate direction. Instead, philosophical reflection "can calmly follow the course of events."[96]

But although this points up a problematic schism between Fichte's hypersystematic conception of the *Wissenschaftslehre* and the actual details of the theory's articulation, it need not be taken to show that the entire Fichtean project has collapsed. In particular, the Deduction of Representation's apparent lack of total rigor in no way impugns the logically prior regressive arguments by which Fichte first isolates the protodiscursive rudiments of world-related subjectivity (the cooperative coordination of which the deduction then attempts to portray). The proper conclusion, then, is not that Fichte's entire undertaking is untenable, but only that his attempt to chart the development of representational awareness out of the dynamic deep structure of the protodiscursive I cannot be carried out successfully within the methodological confines of a strict philosophical foundationalism. If reference to the possibility of experience is strictly forbidden, it seems, then there is no clear way in which a *necessary* deduction of the structures of discursivity could ever get off of its protodiscursive "ground."

NOTES

1. *Johann Gottlieb Fichtes Sämmtliche Werke*, ed. I. H. Fichte (Berlin: Viet & Co., 1845–46), I, p. 222; *Fichte: The Science of Knowledge*, trans. John Lachs and Peter Heath (New York: Appleton-Century-Crofts, 1970), p. 198. *Sämmtliche Werke* was reprinted, along with *Johann Gottlieb Fichtes nachgelassene Werke* (Bonn: Adolphus-Marcus, 1834–35), as *Fichtes Werke* (Berlin: de Gruyter, 1971). Citations of Fichte's writings refer first to the volume and page number in *Sämmtliche Werke*, then to the English translation quoted. In quotations from *Fichte: The Science of Knowledge*, the term "self" has generally been replaced with the more appropriate term "I."

2. *Sämmtliche Werke*, I, p. 15; Daniel Breazeale, ed. and trans., *Fichte: Early Philosophical Works* (Ithaca, N.Y.: Cornell University Press, 1988), p. 69.

3. *Sämmtliche Werke*, I, p. 32; Breazeale, *Fichte*, p. 97.

4. Though the quantity and quality of Fichte scholarship has increased substantially in recent years, the current literature does not (so far as I am aware) include any sustained treatments of the details of this difficult text. Hence the present essay, though indebted to much recent work on Fichte's idealism, will be generally bereft of references to this literature.

5. For more on the importance of Reinhold, in this connection and in general, see Daniel Breazeale, "Between Kant and Fichte: Karl Leonhard Reinhold's 'Elementary Philosophy,' " *Review of Metaphysics* 35 (1982): 785–821. See also: Frederick Beiser, *The Fate of Reason: German Philosophy from Kant to Fichte* (Cambridge: Harvard University Press, 1987), pp. 226–65.

6. Karl Leonhard Reinhold, *Beyträge zur Berichtigung bisheriger Missverständnisse der Philosophen*, vol. 1 (Jena: Mauke, 1790), p. 267.

7. *Sämmtliche Werke*, I, p. 8; Breazeale, *Fichte*, p. 63.

8. *Sämmtliche Werke*, I, p. 334; Breazeale, *Fichte*, p. 246.

9. *Sämmtliche Werke*, I, p. 8; Breazeale, *Fichte*, p. 64.

10. *Sämmtliche Werke* I, pp. 9–10; Breazeale, *Fichte*, p. 65.

11. *Sämmtliche Werke*, I, p. 98; *Fichte: The Science of Knowledge*, p. 99.

12. *Sämmtliche Werke*, I, p. 104; *Fichte: The Science of Knowledge*, p. 104.

13. *Sämmtliche Werke*, I, p. 110; *Fichte: The Science of Knowledge*, p. 110.

14. *Sämmtliche Werke*, I, p. 425; Daniel Breazeale, ed. and trans., *Introductions to the Wissenschaftslehre and Other Writings* (Indianapolis: Hackett, 1994), p. 9.

15. *Sämmtliche Werke*, I, p. 15; Breazeale, *Fichte*, p. 69.

16. *Sämmtliche Werke*, I, pp. 425–26; Breazeale, *Introductions*, pp. 10–11.

17. Ibid.

18. *Sämmtliche Werke*, I, p. 429; Breazeale, *Introductions*, p. 15.

19. *Sämmtliche Werke*, I, pp. 435–42; Breazeale, *Introductions*, pp. 20–27.

20. *Sämmtliche Werke*, I, pp. 227–246; *Fichte: The Science of Knowledge*, pp. 203–17.

21. Owing to space limitations, what follows will be only the most cursory treatment of the pertinent details of Fichte's approach. A more fully developed account of the variety of methods pursued in the *Grundlage* is provided in Daniel Breazeale's paper in this volume.

22. *Sämmtliche Werke*, I, p. 125; *Fichte: The Science of Knowledge*, p. 122.

23. *Sämmtliche Werke*, I, p. 219; *Fichte: The Science of Knowledge*, p. 196.

24. Again, space restrictions require that the niceties of Fichte's doctrine of the *Anstoß* be passed over in silence. A thorough treatment of the issue, carried out with special attention to the *Grundlage*, can be found in Daniel Breazeale, "Check or Checkmate? On the Finitude of the Fichtean Self," in *The Modern Subject: Classical German Idealist Conceptions of the Self*, eds. Karl Ameriks and Dieter Sturma (Albany: State University of New York Press, 1995), pp. 87–114.

25. Ibid.

26. *Sämmtliche Werke*, I, p. 221; *Fichte: The Science of Knowledge*, p. 198.

27. *Sämmtliche Werke*, I, p. 222; *Fichte: The Science of Knowledge*, p. 199.

28. *Sämmtliche Werke*, I, p. 72; Breazeale, *Fichte*, p. 127.

29. *Sämmtliche Werke*, I, p. 71; Breazeale, *Fichte*, p. 126.

30. *Sämmtliche Werke*, I, p. 223; *Fichte: The Science of Knowledge*, p. 199.

31. *Sämmtliche Werke*, I, pp. 224–25; *Fichte: The Science of Knowledge*, pp. 200–201.

32. *Sämmtliche Werke*, I, p. 81; Breazeale, *Fichte*, p. 133.

33. *Sämmtliche Werke*, I, pp. 227–28; *Fichte: The Science of Knowledge*, p. 203.

34. *Sämmtliche Werke*, I, p. 228; *Fichte: The Science of Knowledge*, p. 203.

35. Ibid.

36. Ibid.; but cf. *Sämmtliche Werke*, I, pp. 215f.; *Fichte: The Science of Knowledge*, p. 193.

37. *Sämmtliche Werke*, I, p. 229; *Fichte: The Science of Knowledge*, p. 204.

38. Ibid.

39. Ibid.
40. *Sämmtliche Werke*, I, p. 229; *Fichte: The Science of Knowledge*, p. 204.
41. Ibid.
42. Ibid.
43. *Sämmtliche Werke*, I, p. 230; *Fichte: The Science of Knowledge*, p. 205.
44. Ibid.
45. Ibid.
46. Ibid.
47. *Sämmtliche Werke*, I, p. 231; *Fichte: The Science of Knowledge*, p. 205.
48. *Sämmtliche Werke*, I, p. 231; *Fichte: The Science of Knowledge*, p. 206.
49. *Sämmtliche Werke*, I, p. 232; *Fichte: The Science of Knowledge*, p. 206.
50. Ibid.
51. Ibid.
52. *Sämmtliche Werke*, I, pp. 232–33; *Fichte: The Science of Knowledge*, p. 207.
53. *Sämmtliche Werke*, I, p. 223; *Fichte: The Science of Knowledge*, p. 207.
54. *Sämmtliche Werke*, I, pp. 233–34; *Fichte: The Science of Knowledge*, p. 207.
55. *Sämmtliche Werke*, I, p. 233; *Fichte: The Science of Knowledge*, p. 207.
56. *Sämmtliche Werke*, I, p. 235; *Fichte: The Science of Knowledge*, p. 208.
57. Ibid.
58. *Sämmtliche Werke*, I, p. 235; *Fichte: The Science of Knowledge*, p. 209.
59. See *Sämmtliche Werke*, I, p. 224; *Fichte: The Science of Knowledge*, p. 200.
60. *Sämmtliche Werke*, I, p. 236; *Fichte: The Science of Knowledge*, p. 209.
61. Ibid.
62. *Sämmtliche Werke*, I, p. 236; *Fichte: The Science of Knowledge*, p. 210.
63. *Sämmtliche Werke*, I, p. 237; *Fichte: The Science of Knowledge*, p. 210.
64. Ibid.
65. Ibid.
66. Ibid.
67. Ibid.
68. *Sämmtliche Werke*, I, p. 238; *Fichte: The Science of Knowledge*, p. 211.
69. Ibid.
70. *Sämmtliche Werke*, I, p. 239; *Fichte: The Science of Knowledge*, p. 212. See also *Sämmtliche Werke*, I, p. 212; *Fichte: The Science of Knowledge*, pp. 190f.
71. *Sämmtliche Werke*, I, p. 239; *Fichte: The Science of Knowledge*, p. 212.
72. *Sämmtliche Werke*, I, p. 229; *Fichte: The Science of Knowledge*, p. 204.
73. *Sämmtliche Werke*, I, p. 240; *Fichte: The Science of Knowledge*, p. 212.
74. Ibid.
75. *Sämmtliche Werke*, I, p. 240; *Fichte: The Science of Knowledge*, p. 213.
76. Ibid.
77. Ibid.
78. Ibid.
79. *Sämmtliche Werke*, I, p. 241; *Fichte: The Science of Knowledge*, p. 213.
80. Ibid.

81. *Sämmtliche Werke*, I, p. 241; *Fichte: The Science of Knowledge*, p. 214.
82. Ibid.
83. *Sämmtliche Werke*, I, p. 242; *Fichte: The Science of Knowledge*, p. 214.
84. *Sämmtliche Werke*, I, p. 242; *Fichte: The Science of Knowledge*, pp. 214f.
85. *Sämmtliche Werke*, I, p. 243; *Fichte: The Science of Knowledge*, p. 2115.
86. Ibid.
87. Ibid.
88. Ibid.
89. Ibid.
90. *Sämmtliche Werke*, I, p. 244; *Fichte: The Science of Knowledge*, p. 216.
91. Ibid.
92. Ibid.
93. *Sämmtliche Werke*, I, p. 8; Breazeale, *Fichte*, p. 64.
94. *Sämmtliche Werke*, I, pp. 334–35; *Fichte: The Science of Knowledge*, p. 247.
95. See *Sämmtliche Werke*, I, p. 71; Breazeale, *Fichte*, p. 126.
96. *Sämmtliche Werke*, I, p. 222; *Fichte: The Science of Knowledge*, p. 199.

3

Tom Rockmore

Deduction is a central but still unclarified theme linking together the very disparate philosophers in German idealism. This paper will provide a preliminary account of deduction in Fichte, focusing on his Jena writings, especially the first, or so-called Jena *Wissenschaftslehre*. I will argue that Fichte holds two incompatible views of deduction, one of which is defensible and the other of which is not. I will contend that if our interest in Fichte is not merely historical, we do better to forget about his more Kantian view of deduction and concentrate on his other, rather anti-Kantian view.

DEDUCTION AND FOUNDATIONALISM

The German idealist concern with deduction does not break with but further develops the concern with epistemological foundationalism, which dominates the modern discussion of knowledge, beginning with René Descartes until Husserl and the early Heidegger. Foundationalism is roughly the view that the only way to arrive at knowledge, understood as apodictic, or beyond doubt, hence resistant to skepticism of any kind, is through strictly reasoning away from an initial principal known to be true.[1]

Descartes was a rationalist, who argued from the contents of the mind to indubitable knowledge of the independent external world. In Descartes's wake, as early as Francis Bacon and John Locke a long series of English empiricists maintain that the external world can be directly known. The form of empiricism featured in Locke and other English empiricists is denied by Kant, who rejects the very idea of direct knowledge of the independent external world.

Although he rejects direct knowledge of independent reality, Kant is anything but a skeptic. Like Descartes, like the English empiricists, Kant insists on knowledge beyond skepticism, in his case knowledge from principles that can be rigorously deduced, hence known with certainty. Deduction is the way this strategy is adapted by Kant and the post-Kantian idealists in Germany in order to make similar, foundationalist claims for knowledge. If we cannot have immediate knowledge either through the rationalist inference from the content of the mind to the independent world on which Descartes insists, or through the direct grasp of independent reality as the English empiricists maintain, then we can only have mediate knowledge, or knowledge mediated by a categorial framework.

The main thinkers in this period (Fichte, Schelling and Hegel) were all committed to completing the Kantian critical turning in philosophy, to carrying the Copernican Revolution beyond Kant and through to the end. Although each of them was opposed to the letter of the critical philosohy, they were committed to realizing its spirit.[2] In ways consistent with their respective revisions of the critical philosophy, each of the four main German idealists was concerned to rethink the Kantian conception of deduction, on which Kant relies to make out his claims for knowledge, as part of the task of completing the critical philosophy.

The main post-Kantian philosophers approached the task of developing and completing the critical philosophy in different ways. The ongoing effort to develop the critical philosophy beyond Kant produced a series of misunderstandings of the links between their respective positions and Kant's. Fichte typically proclaimed himself to be the only true Kantian. His claim, which was correctly rejected by Kant, was accepted by the young Schelling and the young Hegel. Until he published the *System of Transcendental Idealism* (1800), Schelling incorrectly understood himself to be a disciple of Fichte. Yet he developed a theory that Fichte correctly regarded as incompatible with his own. In the *Differenzschrift* (1801), his first philosophical publication, Hegel followed Kant's view that there could be no more than a single true philosophical system. Hegel here incorrectly regards Fichte and Schelling as professing variants of the same Kantian theory, although each professed no more than a position that was influenced by but demonstrably different from it. As late as the *Science of Logic*, Hegel continued to emphasize that Kant's theory constitutes the base and starting point of recent German philosophy, including his own.

Some commentators see Hegel's critical attitude toward the critical philosophy as introducing a basic incompatibility between their positions. Certainly, Hegel's theories are incompatible with the letter of Kant's critical philosophy. They are not, however, clearly incompatible with its spirit. In fact, through his acceptance of Fichte's view as the highest form of Kantianism and the one true theory in the *Differenzschrift*, Hegel was committed to developing if not the letter at least the spirit of the critical philosophy. It would, then, be misleading to interpret Hegel's criticisms of it as a basic rejection. Indeed, Hegel explicitly says that the merits of Kant's theory are unaffected by criticism of it.[3]

DEDUCTION IN GERMAN IDEALISM

Deduction is a recurrent theme in the main German idealist texts. In the first *Critique*, Kant offers two transcendental deductions of the pure concepts of the understanding. There are numerous other deductions throughout his writings. Fichte appeals to deduction throughout his writings, most prominently in the Deduction of Representation in the first Jena *Wissenschaftslehre*.[4] Schelling deduces a *System of Transcendental Idealism*, beginning with a deduction of the supreme principle of knowledge.[5] Hegel is often read, especially in the German-language discussion, as the author of a theory culminating in a categorial deduction in the *Science of Logic*.

Post-Kantian German idealists were motivated by their conviction that Kant's deduction of the categories was unsuccessful. Fichte contends that Kant fails to prove that or how representations obtain objective validity since he mistakenly substitutes induction for deduction.[6] Hegel regards the deduction of the categories as authentic idealism and as the speculative principle that Fichte carried further than Kant.[7] He sees Fichte's contribution as in fact deducing the categories Kant only claimed to deduce.

FICHTE ON DEDUCTION

The theme of deduction in Fichte has recently been raised by Ives Radrizzani in his excellent commentary on Fichte's *Foundations of Transcendental Philosophy* (*Wissenschaftslehre nova methodo*, 1796–99). Following Reinhard Lauth and Alexis Philonenko, Radrizzani affirms that "the deduction of intersubjectivity is Fichte's essential problem during the Jena period. . . ."[8]

Fichte is concerned with deduction in many of his writings. In the *Vocation of Man*, written in 1800 at the end of the Jena period, he in effect founds knowledge in faith.[9] In "From a Private Letter," an essay written shortly after he had been discharged from his position in Jena, in a passage on deduction and on those he considers "transcendental" philosophers, Fichte writes:

> I wish to heaven that I had to deal only with such philosophers, and that others might finally understand what the word "deduction" means for me and see that the essence of *my* philosophy, and in my view of *all* genuine philosophy, consists exclusively in deducing![10]

Obviously, Fichte takes deduction as central to his theory; and just as obviously he believes that others have failed to understand his meaning. Yet it is less than obvious how he understands "deduction."

Fichte mentions deduction in many contexts throughout the writings from the Jena period. Here are some of the passages:

In a letter to Reinhold in 1794, he takes a quasi-Cartesian line in maintaining that, like mathematics, philosophy should deduce everything from a single first principle.[11] In his "Outline of the Distinctive Character of the *Wissenschaftslehre*" from 1795, he maintains that deduction establishes facts as facts;[12] and he claims to deduce the basis of all knowing concerned with the products of the acts of the mind and not the acts themselves.[13] In 1797, in the "Annals of Philosophical Tone" he contends that philosophy's task is to provide a derivation of experience as the necessary condition of self-consciousness, but not to deduce air and light.[14] In the same year, in the "Second Introduction to the *Wissenschaftslehre*," he maintains that a deduction is possible only by showing how each category determines the possibility of self-consciousness;[15] and he further affirms in the same text that demonstration is always based on something indemonstrable, so that it yields no more than mediate certainty.[16]

These and other passages are difficult to interpret without placing them in the context of Fichte's wider view. Since Fichte typically insists on a seamless continuity between his theory and Kant's, we can best understand his view of deduction by contrasting it with Kant's.

DEDUCTION IN KANT

The various forms of deduction in Fichte and throughout German idealism are obviously inspired by aspects of Kant's first *Critique*. The notorious problem of the deduction in Kant has given rise to a specialized literature.[17] It will not be possible here to do more than to say a few words about this topic relevant for understanding Fichte's view.

Kant does not invent deduction, which obviously has a long prior heritage in mathematics and philosophy. This theme apparently comes into philosophy in Aristotelian logic. In Kant, the relevance of deduction follows from his conceptually prior, normative view of knowledge as mediated and systematic. Kant famously contends that knowledge necessarily begins in but is not limited to experience. Some empiricists argue for immediate knowledge. For Locke, since simple ideas are directly caused by external things, perceptions in the mind correspond precisely to the nature of external things.[18] Unlike Locke, Kant introduces a distinction between objects of mind that are not given in experience and objects given in experience. For Kant, experience of objects is mediated by the categories, or concepts of the understanding.

Kant insists that knowledge must form a system unified with respect to a single idea.[19] In order to form a unified system, the elements of the system cannot merely be assumed, plucked as it were out of thin air. This would yield no more than a mere rhapsody, as Kant famously remarks with respect to Aristotle. The solution is to deduce the categories, or concepts of the understanding, a task Kant notoriously undertakes in both editions of the *Critique of Pure Reason*.

In the *Critique*, Kant proposes two types of "deduction." In the metaphysical deduction, he proves the a priori origin of categories through agreement with the laws of thought; in the transcendental deduction, he shows their possiblity as a priori modes of knowledge of objects of intuition in general through study of the a priori relation of concepts to object.[20]

Kant's theory of deduction presupposes a legal distinction. By analogy with the juridical question, the *quid juris*, or question of right, as distinguished from the *quid facti*, or question of fact, transcendental deduction considers the justification of the question of right as opposed to the question of fact.[21] This distinction has been extensively discussed. According to H. J. de Vleeschauwer, the *quaestio facti* concerns the demonstration of the existence of a priori intellectual functions, or categories, whereas the *quaestio juris* concerns the justification of the existence of the intellectual functions a priori through a further, higher principle.[22]

DEDUCTION IN FICHTE

The seamless continuity between Fichte and Kant on which Fichte routinely insists is clearly breached with respect to deduction. There is a difference between founding, in the sense of foundationalism, and grounding, a theory. Kant apparently conflates these two senses, which is the significance of his reference to the transcendental deduction as both apodictic and as a mere exposition. In short, while it appears plausible to ground, at present, when foundationalism in any form no longer appears to be a live option, it does not appear plausible to found. It is, hence, more plausible to interpret Kant's *quid juris* in a pragmatic than in an absolute sense, since there is no known way to make out the stronger claim. Kant fails in this attempt, and Fichte does not succeed where Kant fails.

Although Fichte portrays himself as a seamless Kantian, although he is clearly inspired by Kant, and although he may even have believed his assertion that his own theory was only Kant's properly understood, the *Wissenschaftslehre* is finally very different from the critical philosophy. I hold that it is only when we see that, despite his rhetoric, Fichte's theory is his own, not Kant's, from which it basically differs, that we can appreciate the value of his approach that is still worthwhile at a time when we must reject Kant's enterprise.

Fichte follows Kant's emphasis on system, and his view of system as the explanation of experience from the perspective of the subject. Yet the importance of the differences, as could reasonably be expected in the writings of two major thinkers, far outweighs the similarities. An important thinker never simply follows a predecessor but always reformulates earlier concerns in new ways that result in changing the subject as it were. It is a mistake, even when the claim is explicitly made that two thinkers are engaged in precisely the same project, as in Fichte's case, to accept this suggestion without the strictest scrutiny.

Here are three fundamental differences between the two views of deduction. To begin with, Kant studies the conditions of the possibility of objects and knowledge, whereas Fichte focuses on the elucidation of practice, or life, from the metalevel of reflection on life in order to clarify it theoretically. The critical philosophy is a study of the conditions of the possibility of knowledge whatsoever. Kant is concerned, not with how we do know, not with the *quaestio facti* featured in what he refers to as Locke's physiological approach, nor with the physiology of the human understanding. He is rather concerned with an exhaustive deduction of the pure concepts of the understanding prior to and apart from experience as the conditions of experience.

Unlike Kant, Fichte does not focus on the a priori analysis of the conditions of the possibility of experience in general, but rather on the conditions of real experience. He takes experience as a given and argues backward from conditioned to condition thereof in order to explain how experience is really possible. He never attempts to deduce conditions of abstract possibility, or possibility whatsoever. He consistently describes real conditions of actual experience.

This is clear, for instance, in the important account of the fundamental principles of the entire *Wissenschaftslehre* described in the first paragraph of the Jena *Wissenschaftslehre*. Here, building on the insight that we can only provide a theoretical explanation of experience from the perspective of the subject, Fichte characterizes the real interaction of the subject and object of experience instead of arguing from conditions of experience to real experience.

Second, Kant's form of transcendental argument is both prospective in the *Critique of Pure Reason* and retrospective in the *Prolegomena*, where he argues "backward" from conditioned to conditions thereof. In the first *Critique*, Kant claims to make no assumptions about knowledge in order to study its general possibility. In the *Prolegomena*, he assumes straightforwardly that mathematics and natural science provide knowledge. Here, he studies the possibility of mathematics and natural science as well as the conditions of the possibility of the future science of metaphysics that nowhere yet exists. Fichte employs only a retrospective form of transcendental argument. He never argues from an abstract, normative view to the way experience must be.

We need not examine the correctness of Kant's characterization of mathematics that has been sharply criticized by later writers, such as Gottlob Frege[23] and Bertrand Russell.[24] For present purposes, it will be sufficient to to point out that later developments in both logic and geometry clearly undermine the very idea of arguing from something like the conditions of possibility whatsoever to actual experience.

Kant assumes but does not demonstrate that Aristotelian logic provides a list of the operations of the mind. Yet we know through the later development of logic, for instance the logic of relations, that Aristotle does not consider that Aristotelian logic provides no more than a restricted list of the operations of the mind. It follows that Kant's transposition of Aristotle's logical table of judgments as the transcendental table of the categories is indefensible. Similarly, Kant presupposes that there is no more than a single type of geometry to infer from the abstract analysis of space to the conditions of real experience. Yet since we now have non-Euclidean geometries, no inference can be drawn from Euclidean geometry to the nature of experience.

Third, through deduction Kant provides a theoretical foundation of the possibility of knowledge, whereas Fichte provides no more than a practical foundation that, from the strictly Cartesian perspective, is not assimilable to epistemological foundationalism. Descartes takes mathematics, particularly geometry, as the epistemological paradigm. His approach can be characterized as the claim that knowledge, understood as certain knowledge, is possible if and only if the theory is rigorously derived, or deduced from an initial principle that is known to be true.

Like Descartes, Kant is an epistemological foundationalist. In his theory, the transcendental unity of apperception functions in a way similar to the Cartesian cogito. His theory stands or falls on his success in deducing the categories of the understanding. If his deduction fails, then his theory fails. For it then presents, to apply the remark he makes about the alleged failure of Aristotle to deduce the categories of his position, no more than a mere rhapsody that fails the test of rigorous system as he defines it. Deduction is the strategy on which his foundationalism rests.

We can understand the foundationalism of the critical theory by exploring the obvious analogy between Kant and Descartes. Cartesian foundationalism rests on the cogito, whose existence Descartes proves beyond doubt and from which he then rigorously demonstrates the remainder of the theory. If the initial principle is rigorously demonstrated and if the reasoning based on it is rigorous, then the resultant theory is necessarily true.

In the first *Critique*, Kant argues in a similar manner. We can reconstruct his argument as follows. The subject, or "I think" must necesarily be able to accompany all my representations of whatever kind. He clearly hints that his subject is the equivalent of the Cartesian cogito since "I think" or "ich denke" is the German equivalent for Descartes's Latin. He explicitly states that the synthetic unity of the understanding is the highest point of the understanding[25] that, by implication, we can take to be the idea, or principle, under which the other forms of knowledge are united.

It is useful, to understand Kant's argument, to distinguish between a necessary presupposition and a demonstration. We can regard this principle as a mere presupposition that is not proven, since it can follow from no other, prior principle but is rather a necessary condition of representations. In a word, without this principle there can be no representations. Although it functions as the basis the concepts of the understanding, which are deduced through the transcendental deduction, it is not and cannot be demonstrated. Conversely, the categories, which are in principle deduced, hence proven, function as the theoretical foundation of the critical philosophy.

Fichte is often understood as a foundationalist, for instance, in virtue of his interest in system building.[26] A moment's reflection suffices to indicate that this is not foundationalism in the Cartesian theoretical sense that is illustrated in Kant, and probably not foundationalism at all. The Cartesian cogito and the Kantian transcendental unity of apperception are theoretical concepts of subjectivity within an epistemological framework that should not be conflated with a human being or with the human mind. Kant's well-known antipsychologism, which attracted Husserl to the critical philosophy, is inconsistent with any effort to construct a theory on an anthropological basis. In the wake of the great French Revolution, Fichte reinterprets the subject of knowledge and experience as finite human being.[27]

The importance of his reinterpretation of the subject cannot be overemphasized. Unlike Kant, Fichte does not understand the problem of knowledge as that of rational beings whatsoever but rather as knowledge for human beings. He consistently explains experience through the human subject throughout the Jena period before, in the wake of the *Atheismusstreit*, changing his theory in order to explain human being through divine being. Some observers view this change as an important amelioration of the theory. I regard it as the introduction of an important defect that vitiates his thinking after the Jena period. Although the change in Fichte's position brings him closer to Schelling, who consistently invokes God to understand human being, it is in effect a deep mistake to explain finite human being, which is not understood, by infinite being, which is also not understood.

In basing his idea of knowledge on human being, Fichte makes an enormous conceptual break through to the real subject that leads to a theory incompatible with the views of either Kant or Descartes. A theory based on human being is incompatible with Kant's transcendental approach, which rejects psychologism of any kind. Although such a theory can be rigorously systematic, it is also not foundationalist in the Cartesian sense. Descartes reasons strictly on the basis of a theoretical principle that cannot simply be identified with a human subject. For apodictic knowledge is possible for a theoretical, but not for a human, subject. People typically revise their claims for knowledge and, if they are philosophers, their theories of knowledge.

DEDUCTION OF REPRESENTATION IN THE JENA *WISSENSCHAFTSLEHRE*

It is possible that in his zeal to follow Kant, Fichte was also unaware of the distinction between his theory and other forms of foundationalism. We need to distinguish the theory during the Jena period and what he says about it. We can take as our example the initial, or Jena, *Wissenschaftslehre*, the first and most influential of the many versions of his system.

The treatise is divided into three parts, including remarks on the fundamental principles of the entire *Wissenschaftslehre*, followed by exposition of the foundations of theoretical knowledge and then exposition of the foundations of practical knowledge. In his analysis of the fundamental principles (*Grundsätze*) of the entire *Wissenschaftslehre*, Fichte develops Kant's view of the transcendental unity of apperception in three principles: as the assertion of the subject, then of its "negation" by an object defined as that which is not the subject, and finally of the interaction between subject and object. The third basic principle, or the claim that the self and the not-self stand in a relation of reciprocal interdetermination, is analyzed by Fichte as two separate claims that form the content of the theoretical and practical portions of the work.

In the second part, devoted to the foundation (*Grundsatz*) of theoretical knowledge, Fichte explores a view of the subject as determined by the object. In the third part, devoted to the foundation of practical knowledge, he studies the object as determined through the subject. In this book, theory and practice stand in a circular relation. Theory is called forth by practice, and practice is called forth by theory. Yet ultimately, theory is grounded in practice, since we only turn to theory to explain practice when we encounter resistance to our activity in the external world.

The discussion in the theoretical part of the Jena *Wissenschaftslehre* culminates in a detailed summary of the discussion, followed by the Deduction of Representation—not "presentation," as "Vorstellung" is often translated. In the summary, Fichte maintains that we can account for representation through the assumption that the self, or subject, is unlimited. The subject needs to understand the possibility of representation as a fact in its mind through an account of the laws of its own nature.

The discussion in the theoretical part of the work reaches a high point in the Deduction of Representation. Kant's deduction of the categories is motivated through his effort to understand the relation between appearances, or objects given in experience, and what can be thought but does not appear.[28] Fichte's deduction of representation is justified by his later complaint that Kant fails to prove his theory by deduction, since he fails to prove that and how our representations possess objective validity. Fichte's own deduction presupposes inter alia that there is nothing higher than the self;[29] that in philosophy we start from the self that cannot be deduced;[30] and that deduction is a direct, genetic demonstration focused on the self.[31]

Fichte's deduction is a complex argument in no less than eleven steps. Without going through the various steps, suffice it to say informally that, starting from the hypothesis that the self, or subject, is active, Fichte maintains that we can understand the subject as what is left when all objects have been eliminated by the power of abstraction, and the not-self as that from which abstraction can be made. Either can be considered as determined by the other, and conversely. The deduction concludes with the claim that whether the subject is in fact finite, or determined, or infinite, hence determining, in both cases it is reciprocally related merely to itself. According to Fichte, theoretical philosophy can go no further.

CLARIFICATION OF "DEDUCTION" IN KANT

Now that we have reviewed the outline of deduction in Fichte's theory in the Jena period, we need to ask, What is deduction? On one level, the answer is obviously that deduction is a strategy for proof, or epistemological justification, based on a rigorous relation between conditioned and condition. This much is clear. What kind of proof? If we look to Kant, the answer is unclear.

How does Kant understand "deduction"? I believe that his view of deduction is systematically ambiguous. "Deduction" in Kant has at least two distinct, incompatible meanings in the transcendental deduction: "condition of possibility whatsoever" and "exposition [*Darstellung*]."[32] The first meaning is a near synonym for "transcendental argument." Understood in this way, the deduction of the categories uniquely proves the condition of knowledge whatsoever for all rational beings. Through his Copernican turn, Kant straightforwardly claims to elucidate the conditions of the possibility of knowledge whatsoever through what can be characterized, in mathematical language, as an indirect proof. He reasons that, since the only other epistemological strategy is false, and his is at least possible, then it must necessarily be true.

This sense of "deduction" is close to "deduction" in a strictly mathematical sense, with the obvious difference that, unlike mathematics, philosophy claims to avoid presuppositions. The theorems of geometry depend on the claimed truth of the axioms or postulates, which cannot, however, be demonstrated. Such axioms or postulates can be denied without contradiction, as was shown by the denial of Euclid's fifth axiom, the axiom of parallels, early in the nineteenth century in a move that led to the development of non-Euclidean geometries.

The other, weaker sense of "deduction" means "exposition." In this sense, a deduction is no more than a rigorous exposition that, as Kant points out, responds to the *quaestio facti*, or tends to justify in a rigorous, although not an absolute, manner. Here, foundationalism, understood as a method of absolute justification yielding absolute knowledge on a theoretical basis, gives way to a weaker, relative justification yielding no more than relative claims to know, or claims relative to the framework of the theory.

CLARIFICATION OF "DEDUCTION" IN FICHTE

There is a crucial distinction between the elucidation of the conditions of possibility of knowledge whatsoever, an ultimately general claim consistent with the Kantian conception of the nature of transcendental philosophy, and the much weaker assertion (indicated only at the close of Kant's transcendental deduction in the B edition) that deduction comes down to exposition. An exposition is not necessarily an elucidation of the conditions of knowledge whatsoever, although the conditions of knowledge can only be identified through an exposition. This distinction is clear in Kant, but it is not clear to Kant.

This ambiguity is helpful in clarifying "deduction" in Fichte. It is not difficult to detect two views of deduction in Fichte's writings in this period that, for present purposes, we can call the foundationalist and the antifoundationalist or pragmatic interpretations. The key move concerns the status of the initial principle of the deductive chain. For Cartesian foundationalism, including its restatement in the critical philosophy, everything turns on the claim to demonstrate the truth of the initial principle. If not since Aristotle at least since Kant this point separates philosophy from geometry that makes assumptions concerning axioms and postulates that cannot be demonstrated as well as from other, dogmatic philosophical theories that assert what they cannot prove. Other theories can be systematic, but not foundational, if they concede the inability to prove the initial principle that, hence, enjoys the status of a mere hypothesis.

The first, or foundationalist, reading has support in Fichte's writings, both during and particularly after the Jena period, when he appears to provide a religious foundation to the wholly secular theory that initially appears in the Jena *Wissenschaftslehre*. In the "Second Introduction to the Wissenschaftslehre" (1797), he explicitly discusses intellectual intuition. In the next year—in the article "On the Basis of Our Belief in a Divine Governance of the World" that ironically precipitated the *Atheismusstreit* that was to cost him his job and bring the Jena period to a close—he asserts that the highest principle of the *Wissenschaftslehre* cannot be grasped through concepts but can be directly intuited. This suggests that the initial principle underlying deduction functions as a Cartesian epistemological foundation. This inference was shortly drawn by Hegel. In the *Differenzschrift*, he maintains that "[t]he foundation of *Fichte's system* is intellectual intuition, pure thinking of itself, Ego = Ego, I am."[33]

Fichte scholars tend to divide sharply with respect to what I am calling the Hegelian approach to Fichte's position as a form of epistemological foundationalism in the Cartesian traditoin. A generally Hegelian approach to Fichte is accepted by Martial Gueroult,[34] and more recently by Alain Perrinjaquet[35] and Daniel Breazeale,[36] who are in essence committed to a foundational reading of Fichte. This reading of Fichte's position has been refuted by Philonenko,[37] Radrizzani,[38] and myself.[39] Philonenko argues forcefully that, although intellectual intuition is present in the Jena *Wissenschaftslehre*, it does not play the role in this text that Hegel and other foundationalist interpreters of Fichte assign to it. Radrizzani maintains that the inability to demonstrate the initial principle of his position commits Fichte to a circular approach in which the foundation of the system is also founded by the system following from it.[40] Elsewhere I have argued that Hegel saw this consequence and made it the basis of his own circular approach to knowledge.[41]

In order for intellectual intuition of the self to be the fundamental principle, it must be the case that Fichte bases his theory on it and that he knows this principle to be true. Yet clearly the concept of the self as Fichte understands it is a hypothesis that cannot be deduced, but is merely assumed as a basis for discussion. Fichte insists that philosophy should deduce everything from a first principle; but it would be hasty to conclude that he thinks he has done this or even that it is possible to do so. He clearly contends that an absolutely initial principle, such as a Cartesian foundation, cannot be proven or defined.[42] For since such a principle depends on no prior principle, it cannot be deduced from anything else.

Fichte's theory, which is based on a conception of the human subject, is not epistemological foundationalism, since it lacks a Cartesian foundation. What does he mean by "deduction"? The alternative to epistemological foundationalism can only be antifoundationalism of some undefined kind. Here two passages are helpful. In the first Jena *Wissenschaftslehre*, in the middle of his deduction of presentation, Fichte makes the surprising statement that "[t]he *Wissenschaftslehre* is to be a pragmatic history of the human mind."[43] In an important metaphilosophical text from the same year, "Concerning the Concept of the *Wissenschaftslehre*," written while he was working out the Jena *Wissenschaftslehre*, he writes: "We are not the legislators of the human mind, but rather its historians. We are not, of course journalists, but rather writers of pragmatic history."[44]

How can we interpret these passages? It has been suggested through reference to Kant's remark in the "Fundamental Principles of the Metaphysics of Morals" that a pragmatic history teaches prudence. Since Fichte was anything but prudent, this suggestion seems to me be improbable. It is probable that Fichte is also not alluding to the passage in the *Critique of Pure Reason*, where Kant remarks that philosophers are legislators of human reason.[45] I think that Fichte rather means to use "pragmatic" in an epistemological sense in both passages to indicate that in the final analysis philosophical theories, including his own, are at best probable and never certain since they are unable to demonstrate their basic premises.

The evidence for this reading is contained in a number of explicit passages where Fichte clearly and forcefully denies that philosophy can ever be the kind of epistemologically seamless web that Descartes, Kant and others have long sought to weave where there are no assumptions since everything is demonstrated. It is at least arguable that theories cannot prove everything since they rest on assumptions that cannot themselves be demonstrated through the theory based on them.

There are many indications that Fichte holds this view in the Jena period. In section 1 of the Jena *Wissenschaftslehre*, he remarks on the need to presuppose the laws of logic that will later be derived through a circular argument.[46] Yet circular arguments do not yield certainty. In the same passage, he refers to section 7 in his metaphilosophical text, where he says in reference to his theory that it is probable only and can never be proven in strict fashion; for one should never claim infallibity.[47] Again in section 1 of the Jena *Wissenschaftslehre*, he indicates the need to start from a fact from which the existence of the subject cannot be deduced.[48] In section 9 of the Jena *Wissenschaftslehre*, he indicates the need to start from the self as subject that cannot be deduced.[49]

Fichte's view of deduction is often, in fact routinely, understood in terms of Kant's position. Yet these and other passages that could be cited suggest a very different reading of Fichte's view of deduction in Fichte in the Jena period. The term "pragmatic history," which recurs twice in the Jena writings, is well chosen. Rather than a deduction of the conditions of possibility whatsoever, that is, deduction in anything like a Kantian sense, what Fichte presents is an exposition, as he points out, a likely story that can never be strictly proven, based on certain assumptions from which he reasons rigorously.

Descartes, who made fundamental contributions to geometry, suggests through his conception of epistemoligical foundationalism that philosophy can and indeed must surpass mathematics. On the contrary, for Fichte philosophy is like geometry in reasoning rigorously on the basis of presuppositions that it does not and cannot establish. Like mathematics, philosophy remains certain, but only within the framework it presupposes.

CONCLUSION

I come now to the conclusion. Original philosophers, like Fichte, typically bring preceding thought to a new, higher level and innovate in introducing new ideas. Since Hegel, Fichte has often been read in a foundationalist manner as providing a theory that necessarily describes the conditions of knowledge. There is some evidence that he holds this view of deduction. Yet there is also compelling evidence that he simultaneously holds a rather different, more interesting view according to which a philosopher can provide no more than a likely story, a distant successor of Plato's myth.

If this is Fichte's view of deduction, then it is eminently defensible. The very idea of the elucidation of the only possibility of knowledge whatsoever that underlies Kantian deduction has an antihistorical ring. Kant's clearly mistaken assumption that Euclidean geometry would stand forever as the only a priori analysis of space undermines his own philosophical position and illustrates the real difficulty of identifying substantive claims unrelated to their historical moment. Although deduction can be rigorous, I believe it can never yield apodictic knowledge. The view that it in fact does provide certain knowledge rests on a confusion between first order claims to know, such as the familiar illustration that the cat is on the mat, and second order claims about knowledge of the conditions of knowledge, a confusion that is already present in Kant.

Fichte is himself prey to this confusion, since he appears to hold two conflicting views. On the one hand, he argues that proof always rests on an asumption or assumptions that cannot be proven; and, on the other, he insists we have an intuitive grasp of that which cannot be known conceptually. In a word, he conflates foundationalist and antifoundationalist, or linear and circular, forms of proof. Yet it is the latter that is correct. For all theory of whatever kind never yields anything more than a possible perspective. As Fichte points out, deduction provides no more than a pragmatic history of the human mind, in short a likely story.

NOTES

1. This is obviously what Descartes does. He has many followers, including Chisholm. See "A Version of Foundationalism," in Roderick M. Chisholm, *The Foundations of Knowing* (Minneapolis: University of Minnesota Press, 1982), pp. 3–32.

2. See Tom Rockmore, *Hegel's Circular Epistemology* (Bloomington, Ind.: Indiana University Press, 1986), *Before and After Hegel: A Historical Introduction to Hegel's Theory* (Berkeley: University of California Press, 1993), *On Hegel's Epistemology and Contemporary Philosophy* (Amherst, N.Y.: Humanity Books, 1995).

3. See G. W. F. Hegel, *Science of Logic*, trans. A. V. Miller (Atlantic Highlands, N.J.: Humanities Press, 1969), p. 61 n. 1.

4. See F. G. Fichte, *The Science of Knowledge*, trans. Peter Heath and John Lachs (Cambridge: Cambridge University Press, 1982), pp. 203–18.

5. See F. W. J. Schelling, *System of Transcendental Idealism*, trans. Peter Heath, with an introduction by Michael Vater (Charlottesville: University of Virginia Press, 1978), pp. 24–31.

6. See *FICHTE: Foundations of Transcendental Philosophy (Wissenschaftslehre) nova methodo (1796–99)*, ed. and trans. Daniel Breazeale (Ithaca: Cornell University Press, 1992), p. 80.

7. See G. W. F. Hegel, *The Difference Between Fichte's and Schelling's System of Philosophy*, trans. H. S. Harris and Walter Cerf (Albany: State University of New York Press, 1977), p. 79.

8. Ives Radrizzani, *Vers la Fondation de l'intersubjectivité chez Fichte, Des Principes à la Nova Methodo* (Paris: Vrin, 1993), p. 40.

9. See J. G. Fichte, *The Vocation of Man*, ed. Roderick M. Chisholm (Indianapolis: LLA, 1950).

10. "From a Private Letter," in J. G. Fichte, *Introductions to the Wissenschaftslehre and Other Writings*, ed. and trans. Daniel Breazeale (Indianapolis: Hackett, 1994), p. 175.

11. See Letter to Reinhold, dated January 15, 1794, in *Fichte: Early Philosophical Writings*, ed. and trans. Daniel Breazeale (Ithaca, N.Y.: Cornell University Press, 1998), p. 373.

12. See *Fichte: Early Philosophical Writings*, p. 247.

13. See ibid., p. 255.

14. See ibid., p. 347.

15. See Fichte, *Introductions to the Wissenschaftslehre*, p. 63.

16. See ibid., p. 93.

17. For a useful overview, see H. J. de Vleeschauwer, *La Déduction transcendentale dans l'oeuvre de Kant* (Librairie Ernest Leroux, 1935–1937), 3 vols., especially vol. 2, pp. 143–91.

18. See John Locke, *An Essay Concerning Human Understanding*, ed. Alexander Campbell Fraser (New York: Dover, 1959), vol. 1, p. 521.

19. Immanuel Kant, *Critique of Pure Reason*, trans. Norman Kemp Smith (London/New York: Macmillan/St. Martin's, 1961), p. 653.

20. See ibid., B159, p. 170.

21. See ibid., B116, p. 120.

22. See Vleeschauwer, *La Déduction transcendentale dans l'oeuvre de Kant*, vol. 2, pp. 144–45.

23. See Gottlob Frege, *The Foundations of Arithmetic*, translated by J. L. Austin (New York: Harper and Brothers, 1950).

24. See Bertrand Russell, *An Essay on the Foundations of Geometry* (New York: Dover, 1956).

25. See Kant, *Critique of Pure Reason*, B134, p. 154.

26. See the articles by Breazeale and Perrinjaquet in *Fichte: Historical Context and Contemporary Controversies*, ed. Daniel Breazeale and Tom Rockmore (Atlantic Highlands, NJ: Humanities Press, 1994); see also the editor's introduction to Fichte, *Introductions to the Wissenschaftslehre*, p. viii.

27. See Tom Rockmore, "Fichte, die subjektive Wende, und der kartesianische Traum," in *Fichte-Studien*, Band 9 (Atlanta: Rodopi, 1997), pp. 115–26.

28. See Kant, *Critique of Pure Reason*, Bxxvii, p. 27.

29. See Fichte, *Science of Knowledge*, p. 224.

30. See ibid., p. 262.

31. See ibid., pp. 239, 269.

32. See Kant, *Critique of Pure Reason*, B168, p. 175.

33. Hegel, *The Difference Between Fichte's and Hegel's System of Philosophy*, p. 118.

34. See Martial Gueroult, *L'évolution et la structure de la doctrine de la science chez Fichte* (Paris: Aubier-Montaigne, 1930).

35. See Alain Perrinjaquet, "Some Remarks Concerning the Circularity of Philosophy and the Evidence of Its First Principle in the Jena *Wissenschaftslehre*," in *Fichte: Historical Contexts/Contemporary Controversies*, pp. 71–95.

36. See Daniel Breazeale, "Circles and Grounds in the Jena *Wissenschaftslehre*," in *Fichte: Historical Contexts/Contemporary Controversies*, pp. 43–70; see also his editor's introduction to Fichte, *Introductions to the Wissenschaftslehre and Other Writings*, p. viii.

37. See Alexis Philonenko, *La Liberté humaine dans la philosophie de Fichte* (Paris:Vrin, 1966).

38. See Radrizzani, *Vers la Fondation de l'intersubjectivité chez Fichte*, pp. 56–58.

39. See Tom Rockmore, "Antifoundationalism, Cricularity, and the Spirit of Fichte," in *Fichte: Historical Contexts/Contemporary Controversies*, pp. 96–112; see also Tom Rockmore, "Fichtean Circularity, Antifoundationalism, and Groundless System," *Idealistic Studies* 25, no. 1 (1995): 107–24.

40. See Radrizzani, *Vers la Fondation de l'intersubjectivité chez Fichte*, p. 67.

41. See Rockmore, *Hegel's Circular Epistemology*.

42. See Fichte, *Science of Knowledge*, p. 93.

43. Ibid., pp. 198–99; translation modified.

44. *Fichte: Early Philosophical Writings*, p. 131.

45. See Kant, *Critique of Pure Reason*, B868, p. 658.

46. See Fichte, *Science of Knowledge*, pp. 93–94.

47. See *Fichte: Early Philosophical Writings*, p. 130.

48. See Fichte, *Science of Knowledge*, p. 99.

49. Ibid., p. 262.

SPECIAL
ISSUES IN THE
GRUNDLAGE

Part 3

4

SELF-MEASURE AND SELF-MODERATION IN FICHTE'S *WISSENSCHAFTSLEHRE*

Michael Baur

INTRODUCTION

In the opening chapter of his *Essay Concerning Human Understanding*, John Locke explains that the self-understanding or self-measure of the human mind includes an account of the mind's limits, and so the mind's self-understanding can provide adequate grounds for intellectual self-moderation or self-control: "If we can find out, how far the Understanding can extend its view; how far it has Faculties to attain Certainty; and in what Cases it can only judge and guess, we may learn to content our selves with what is attainable by us in this State."[1] Furthermore: "If we can find out those Measures, whereby a rational Creature put in that State, which Man is in, in this World, may, and ought to govern his Opinions, and Actions depending thereon, we need not be troubled, that some other things escape our Knowledge."[2]

Compared to Locke's *Essay Concerning Human Understanding*, Fichte's *Wissenschaftslehre* may appear to exemplify the very opposite of intellectual modesty and self-control. Unlike Locke, Fichte argues that a true system of knowledge should not seek to limit or moderate itself by reference to what is allegedly unknowable outside of it. A true system of knowledge, writes Fichte, ". . . only has to agree with itself. It can be explained only by itself, and it can be proven—or refuted—only on its own terms."[3] Furthermore, Fichte suggests that an appreciation of his system of knowledge requires not modesty in his readers, but rather an implicit sense of superiority: "I wish to have nothing to do with those who, as a result of protracted spiritual servitude, have lost their own selves and, along with this loss of themselves, have lost any feeling for their own conviction. . . ."[4]

In this essay, I shall seek to show that, contrary to initial impressions, it is Fichtean idealism, and not Lockean (or any similar) realism, that is truly modest.[5] According to my account, Locke's explicit declarations concerning the modesty and limitedness of his own project exhibit a certain ignorance concerning the genuine problems at issue. In his claim to be modest, the realist Locke professes to know more than he actually does, and thus manifests his own immodesty. By contrast, the idealist Fichte must refrain from such direct claims to modesty and must *appear* to be immodest, precisely because he has a greater understanding of how *radically* limited human knowledge always is. Like Plato's Charmides, Fichte realizes that any explicit claim to self-limitation and self-moderation would actually give lie to itself.[6]

I. THE PROBLEMS AND PARADOXES OF SELF-MEASURE

As my reference to Plato suggests, Fichte's thought concerning the problems and paradoxes of self-measure can be situated within an extended and rich philosophical tradition. And so our consideration of some of these issues might well begin with a consideration of Plato. In book 4 of Plato's *Republic*, Socrates suggests that genuine self-moderation or self-control is impossible, and thus perhaps the very idea of self-moderation or self-control should be dismissed as "ridiculous."[7] After all, a self that is in need of control would have to be a self that is unruly or undisciplined in some way; however, a self that is unruly or undisciplined would be lacking precisely what would be needed for *self*-control. On the other hand, a self that is *capable* of controlling itself would have to contain some principle of discipline or control *within* itself; but such a self would then not *need* to be controlled or disciplined, and thus any activity that the self happened to exercise upon itself would not be real self-*control*, but some other form of self-relation. Thus where control is really *needed*, *self*-control is impossible; and where *self*-control seems to be possible, *control* is not really *needed*, and thus self-*control* is not possible. Socrates does acknowledge that one can talk of "self-control" in an incidental sense; however, a close examination of the issue will always reveal that what can be controlled by the self, precisely by virtue of its *need* to be controlled, is never the self qua self, but something *other* than what is doing the controlling. One can thus never say that the self qua self is controlling itself; at most, one can say only that one *part* of the self is controlling some *other* part.

The problems and paradoxes of self-control reappear in slightly different form when one considers the issue of the mind's (or the rational self's) measuring or testing of its own knowledge.[8] First of all, genuine self-measure or self-testing is not possible for a mind that is infallible. Genuine self-*measure* or self-*testing* presupposes at least the possibility of correction, and thus the possibility of error; for without the possibility of correction, a mind's relating to itself might be described as a kind of self-agreement or self-affirmation, but not as *self*-measure or *self*-testing. Furthermore, genuine self-measure or self-testing is not possible for a mind that is fallible but does not know itself as fallible. A fallible mind that does not know itself as fallible is indeed capable of being measured or tested, but only by someone or something *other* than itself. Genuine *self*-measure or *self*-testing requires that the self in question can at least conceive the possibility of its own being corrected, and thus the possibility of its own being in error; the testing or measure of a mind that does not know itself as fallible cannot be a *self*-testing or *self*-measure. It would seem, then, that *self-testing* or *self-measure* is genuinely possible only for a mind that is fallible, and that knows itself as fallible. But even this seems to be an impossibility.

If a mind is fallible, then there is in principle no reason why its fallibility does not extend to any attempted act of self-measurement or self-testing. In other words, there is nothing to rule out the possibility that any determinate standard to which the fallible mind might appeal in its act of self-measurement might be invalid or mistaken. Such a mind, then, is not merely fallible, but *radically* fallible; its fallibility extends in principle to any attempted act of self-measurement. Of course, one can suggest that the fallible mind might somehow hit up upon the right standard for itself; however, the fallible mind could never demonstrate *for itself* that such is indeed the right standard for measuring itself. For any such act of demonstrating is susceptible to the same kind of fallibility which pertains to the fallible mind as fallible. Accordingly, the validity of the standard for measuring or testing the mind can never be demonstrated by or for the mind to be tested, but only by or for a mind *other* than the mind to be measured or tested.[9]

We thus have what appears to be an insoluble impasse concerning the issue of intellectual self-measure. On the one hand, in order to have genuine self-*measure*, that by which the fallible mind is to be measured (the standard or criterion) must be *other* than the mind itself; if it is not *other* than the mind itself, then we do not have genuine self-*measure*, but rather simple self-relation or self-agreement. On the other hand, precisely because the mind to be measured is necessarily fallible (for it would make no sense to measure or test an infallible mind), there can be no guarantee that the standard by which the mind chooses to measure itself is itself not mistaken or misapplied. We can thus state the problem in general terms: all genuine *measurement* requires an appeal to something *other* than what is being measured, but in intellectual "*self*-measure" the other is never really a genuine other, but only an other as it is understood or applied (and thus perhaps misunderstood or misapplied) by the (fallible) rational self.

If the problem of intellectual self-measure is simply accepted as it has been formulated thus far, then the problem would indeed appear to be insoluble. According to my account, Fichte's solution to the problem consists in his disruption of our fixation on the mere *formulation* of the problem and his directing of our attention back upon our *awareness* of the problem *as* a problem. Like Socrates, Fichte aims to get us to reflect not so much on the theoretical problem as it stands before us, but rather on our *activity* and *involvement* in being puzzled by the problem in the first place. Through his *Wissenschaftslehre*, Fichte suggests that it is *our awareness* of the problem as a problem that constitutes the beginning of a genuine solution to it.

What is at stake in the problem of intellectual self-measure is nothing less than the possibility of philosophy, if philosophy is understood to be a form of rational discourse that aims to test or justify rational discourse itself. After all, any attempt by philosophy to test or justify rational discourse must take place within the medium of rational discourse itself. Accordingly, the kind of rational discourse that is enacted in philosophy is a species of the self-testing or self-measuring that is our primary concern in this essay. If it should turn out that intellectual self-measure is impossible in general, then philosophy itself is also impossible.

Fichte acknowledges that the activity of intellectual self-measure as enacted through philosophy is inevitably circular insofar as one cannot demonstrate that such self-measure is possible without engaging in the activity itself. The point is not to escape the circularity, but to engage in it self-consciously. Furthermore, one should not be misled into thinking that the circularity alone amounts to a demonstration of the impossibility of philosophy. The circularity cuts both ways: the possibility of philosophy can be demonstrated only through its actuality; but conversely, the mere nonactuality of a self-justifying philosophical system does not amount to a valid proof of its impossibility. One can demonstrate neither the *possibility* nor the *impossibility* of philosophical self-measure from some external point of view. "The question concerning the possibility of philosophy is thus itself a philosophical question."[10]

II. What Is Implicit in Our Awareness of the Problem

Fichte's *Wissenschaftslehre* can be understood as an invitation to enter into an activity that will ultimately demonstrate that our *awareness* of the problem of intellectual self-measure *as* a problem constitutes the beginning of a solution to the problem. But what is entailed in our *awareness* of the problem as a problem?

It was suggested above that the problem of intellectual self-measure arises as a genuine problem for us because we are aware of our radical fallibility as knowers. Furthermore, the human mind is radically fallible to the extent that its fallibility extends in principle to *any* attempted act of self-measurement and *any* attempted act of demonstrating the validity of a given standard for such measurement. To recognize that the mind is radically fallible is to recognize that no given content or standard is necessarily determinative for the self's thinking. In turn, to be aware that no given content or standard is *necessarily* determinative for the self's thinking is to be aware that the self's thinking is not determined by any external necessity, but is radically free. The *meaning* of this freedom is susceptible to further elaboration; for now it need not mean anything more than that no given content or standard (or idea or representation) necessarily imposes itself on us and forces or causes us knowers to accept it as true.

The important point is not merely that we can be wrong about any given content or standard that presents itself to us. What is important is that we are *self-consciously aware* that we can always be wrong about any given content or standard. The self's awareness of the *radical fallibility* of its own knowing thus coincides with the self's awareness of its own *radical freedom*. Stated differently, the self's own self-conscious awareness of the radical fallibility of its own knowing is possible only as an enactment of the self's radical freedom. We can be aware of our radical fallibility as knowers only because we are implicitly aware that we are radically free (i.e., that no given content or standard is necessarily determinative for our thinking).

Broadly understood, the goal of Fichte's *Wissenschaftslehre* is to begin with such skepticism about theoretical knowing and to develop a system by articulating what is implicit in such skepticism. For Fichte, as for Socrates, what is crucial is not the mere *fact* that we are radically fallible (ignorant), but rather *our activity in being aware* that we are radically fallible (ignorant). When our fallibility (or ignorance) becomes *self-conscious* fallibility (or ignorance), the result is not mere emptiness, but rather a species of wisdom (or system).

Fichte explicitly acknowledges that skepticism about our capacity to know has always been integral to the achievements of systematic philosophy: "It is undeniable that philosophizing reason owes all the human progress that is has made so far to the observations of skepticism concerning the insecurity of every resting place yet obtained by reason."[11] It is unfortunate that the quasi-deductive structure of Fichte's *Grundlage der gesamten Wissenschaftslehre* has misled many readers into thinking that Fichte's systematic philosophy is entirely different from the movement of self-conscious fallibility. As I shall try to show, Fichte's entire *Grundlage* can be understood as the self-articulation of human fallibility becoming fully self-conscious.

Also implicit in our awareness of the problem of intellectual self-measure is the inadequacy of realism as a solution to the problem. Stated in terms of the problem of self-measure, "realism" refers to any form of knowing that seeks to measure the adequacy of itself as a form of knowing by appealing to some content or standard that is supposed to exist entirely independent of itself. According to Fichte, a careful analysis of the *problem* itself will reveal two fundamental weaknesses in the realistic solution to it.

First of all, the realist's way of thinking is fundamentally arrogant or immodest. On the face of it, the "realistic" way of thinking may appear to be more modest than Fichte's idealistic way of thinking. After all, the realist argues that there are things that are simply independent of the human mind, things that necessarily place limits upon what we can truthfully think. For the realist, these external things constitute the proper standard for testing the validity of human knowledge, and it would be arrogant to refuse to conform one's thought to these external things.

In spite of initial appearances, the realist's claim to moderation amounts to a kind of immodesty that masks its own immodest character. The realist, after all, is not merely claiming that the human mind is fallible (a claim with which the Fichtean idealist would agree). The realist is also making a claim about the *ground* of the mind's fallibility: the mind is fallible because its own knowing may not always measure up to external things, which constitute the proper criterion or standard for testing the mind's knowing. Implicit in this claim is the realist's (immodest) belief that it is possible, in principle, for the fallible human mind to escape its own fallibility and attain knowledge of external things as they exist in themselves, entirely unconditioned by the activity of the fallible mind itself. In other words, the realist implicitly claims to be able to achieve a God's-eye view of the self-in-itself and the thing-in-itself, as well as the relation of similarity (or difference) that holds between the two. The realist's claim thus amounts to an infinitizing of consciousness.

For Fichte, one who seeks to affirm the genuine fallibility of human knowing must refrain from accounting for such fallibility by appealing to some kind of determinate reality that can putatively be known to exist as independent of and unconditioned by the finite rational self's own way of knowing. Paradoxically, Fichte's idealistic claim that the rational self does not ultimately measure itself by anything that exists independently of itself does not amount to an infinitizing of consciousness. Fichte's idealism is not a sign of arrogance; it is a sign that Fichte has recognized and accepted the radical fallibility of human knowing all the more genuinely.

In addition to being fundamentally immodest, the realist's way of thinking, according to Fichte, is also implicitly self-contradictory. As we have seen, the problem of self-measure arises as a genuine problem because the fallible mind recognizes the radicalness of its own fallibility; the problem arises insofar as the rational self recognizes that no given content or standard is necessarily determinative for its own thinking. Nevertheless, in searching for a given content or standard by which to measure the self's thinking, the realist is, in effect, searching for a content or standard that is supposed to be necessarily determinative for the self's thinking. After all, if the content or standard sought by the realist were genuinely independent of consciousness itself, then it would have to be, qua standard, unconditioned by the activity of the fallible rational self being measured. In other words, a standard that is genuinely independent of consciousness itself would have to be unaffected, qua standard, by any change or distortion that consciousness might introduce as consciousness applies (or misapplies) the standard to itself in its act of self-measuring. But if the external standard is thus unchanged and unaffected, qua standard, by the self's interpretive activity (i.e., if the standard remains valid for thought, no matter what the self might make of it), then that standard would have to be, qua standard, necessarily determinative for the consciousness to be tested. (As soon as the standard is *not* thus immune to being shaped, conditioned, or interpreted by the self that applies it in the act of self-testing, the standard is no longer genuinely independent and unconditioned.) In short, the realist's solution to the problem of self-measure posits the existence of some standard or content that is supposed to be necessarily determinative for thought; however, the solution that the realist proposes implicitly contradicts the self-conscious fallibility that gave rise to the problem in the first place.

We must keep in mind that the implicit contradiction in the realist's way of thinking is recognized as a contradiction by the observing idealist, and not by those realists who remain trapped within the contradiction. Indeed, the contradiction *cannot* be recognized by the realist, who has not fully appreciated the radicalness of human fallibility and freedom. Along these lines, Fichte acknowledges that his own idealistic position must be "dogmatically opposed" to realism, as well as to any position that even holds open the possibility of realism as a viable system.[12]

On the face of it, Fichte's claim concerning the necessity of this "dogmatic opposition" might appear to be a form of arrogance and immodesty. It might seem that a more modest approach would require Fichte to seek some kind of *rapprochement* between his own position and that of the realist (or the position of someone who holds open the possibility of realism). But just the opposite is the case: any proposed *rapprochement* between the two opposed positions would presuppose that there is some third, independent thing to which the realist and the idealist would both have access, and to which they could appeal in order to settle their differences. In other words, the proposal for a *rapprochement* between realism and idealism is itself based on the realist's immodest bias. By contrast, Fichte's idealism explicitly recognizes the inability of the fallible rational self to know any determinate "third thing" or standard as it might be "in itself," unrelated to and unconditioned by the activity of the fallible self. Thus Fichte's insistence on the necessity of his dogmatic opposition to realism is not a result of arrogance or immodesty, but is rather a corollary to his awareness of the radical limits of his own thinking and ability to persuade.

By acknowledging that his own position must be dogmatically opposed to that of the realist, Fichte is affirming that there is no possibility of persuading or coercing his critics to see things as he sees them, if they do not *already* acknowledge their own radical fallibility and freedom. Stated differently, Fichte is saying that the activity of self-conscious fallibility and freedom that underlies his own idealism cannot be induced or forced upon others through any kind of mediation, explanation, or argumentation. Philosophical discourse must simply *begin* with the activity of free, self-conscious fallibility, the hidden source of all wonder. Any force or any appeal to some external "third thing" contradicts the essence of that freedom which lies at the basis of one's self-conscious fallibility.

The point here is not that Fichte's idealism must be dogmatically opposed to realism, while realism is not dogmatically opposed to idealism. Idealism and realism are each dogmatically opposed to one another; but the ways in which they are so opposed is fundamentally different. Fichte's idealistic position is dogmatically opposed to realism in a way that explicitly includes an awareness of its own dogmatic opposition. By contrast, the realist position is dogmatically opposed to the idealist position, but in a manner that seeks to deny the necessity of its own dogmatic opposition: realism steadfastly refuses to accept its own limits and thus continually seeks a common point of agreement between itself and idealism (by appealing to some thing-in-itself outside of all knowing). To this extent, Fichte is like Socrates, and the realist is like the poets, politicians, and craftsmen. Both are necessarily ignorant of the "third thing" outside of knowing by which they might finally settle their differences. But while Socrates (Fichte) acknowledges his own ignorance, the poets, politicians, and craftsmen (the realists) do not.

III. SELF AND NOT-SELF

Our understanding of the radicalness of the problem of intellectual self-measure implicitly includes within it an awareness of our radical fallibility and freedom as knowers. In order to show how our awareness of the problem constitutes the beginning of a solution to it, we must say more about the rational self that is aware of itself as radically fallible and free.

To be aware of oneself as radically fallible and free is to be aware that no given content is necessarily determinative for one's thinking. But exactly how are we to understand and define the rational self that is thus self-aware? Any attempted definition of the rational self as self-consciously fallible and free cannot be based on or derived from any given content. After all, the self's radical fallibility extends to any proposed definition of the self that is based on or derived from any given content (or idea or representation). If we are to remain sensitive to the problem of our radical fallibility, then—in our search for a solution to the problem—we must exercise extreme skeptical restraint in our definition of the self: in defining the self, we must refrain entirely from relying on any given content or idea or representation.

This skeptical restraint, however, does not prevent us from defining the self-consciously fallible, free self. Indeed, it is this very restraint that gives us just what we need in order to define the rational self without recourse to any given content. The rational self is nothing other than the activity of being aware of itself as radically fallible and free, undetermined by any given content. If we are to take the problem of self-measure seriously, we *must* define the self in this way. Any suggestion that the self might be more adequately defined by reference to something *other* than such activity implicitly involves the problematic, realistic claim that the self can have knowledge of some external thing as it is in itself, independent of and unconditioned by the self's own activity of being aware of itself.

The awareness of oneself as radically fallible and free, an awareness that constitutes the self's very being, is necessarily a *nonrepresentational* kind of awareness. Any given representation inevitably belongs to that sphere of given contents to which one may not appeal in defining the self. Thus the term "awareness" must be used here with caution: this awareness is nothing like any empirical awareness of a given, determinate content (derived either from internal or external sense). The kind of awareness that constitutes the self's being does not refer to or depend on any given content or fact (*Tatsache*) whatsoever, but is simply an activity (*Tathandlung*), namely the activity of being aware, in a nonrepresentational way, of oneself as free and undetermined by any given content.

With this, we have arrived at the first principle of Fichte's *Grundlage der gesamten Wissenschaftslehre*, the pure *Ich = Ich*. This activity is alternatively described by Fichte as the activity of self-positing, or the activity of simple "being *for* self." The "content" of the first principle of the *Grundlage* is thus nothing other than the activity of self-positing, or being *for* oneself in a nonrepresentational way. Here, the act of self-awareness and the content of the act fully coincide; all that the self is, is simply its own act of being for self, and all that is for the self, is simply its own selfhood as the act of being for self: "*To posit oneself* and *to be* are, as applied to the self, perfectly identical. Thus the proposition, 'I am, because I have posited myself' can also be stated as: '*I am absolutely [schlechthin], because I am.*' "[13]

Fichte's description of the self-positing self as "absolute" (*absolut*), and the translation of the German "*schlechthin*" as "absolute" or "absolutely," can be misleading. Saying that the rational self "absolutely" posits itself is not an attempt to infinitize the self, but rather an attempt to express the radicalness of the self's fallibility. To say that the self "simply" or "absolutely" posits itself is to say that the self is so radically fallible as a knower that it is absolutely unable to explain itself or (what amounts to the same thing) explain its awareness of itself by appealing to any thing other than itself. It would be fundamentally immodest for the self to try to explain itself by appealing to an external state of affairs that somehow preexisted the self and "caused" the self to become the self-consciously fallible self that it is. Any such explanation would involve the self in the immodest claim that it can, as a fallible self, have knowledge of a state of affairs that is independent of itself, unrelated to and unconditioned by the self's own fallible activity. The self must simply *begin* with itself as self-consciously fallible and free, and it can never get "behind" or "ahead" of this starting point by appealing to any kind of condition, causation, or mediation. Far from infinitizing the self, Fichte's discussion of the self as simply (*schlechthin*) self-positing is a much more thorough and honest admission of the self's radical fallibility.[14]

Because of the skeptical restraint that we must exercise in defining the self, we must also refrain from thinking of the self as any kind of substance or thing at all. The rational self is not a thing that also happens to think (a *res cogitans*); it is nothing but the activity of thinking. The rational self ". . . is an *act*, and absolutely [*absolut*] nothing more; we should not even call it an *active* something [*ein Thätiges*]. . . ."[15] The rational self is nothing other than the "pure activity" of nonrepresentational, nonsubstantialist self-awareness. Any proposed definition of the self as an active *something* or active *substance* implicitly refers to some *being* or *substance* that can allegedly be known to exist as it is in itself, apart from the bare activity of the self's thinking. Thus any substantialist or reifying vision of the self shares the realist's immodest bias. Again, the exclusion of all passivity in this definition of the self is not an attempt to infinitize the self, but rather an attempt to acknowledge the self's radical fallibility more honestly and completely.

This account of Fichte's first principle in the *Grundlage* sheds some light on the question of whether Fichte's *Wissenschaftslehre* should be understood as a form of foundationalism or antifoundationalism. According to my account, Fichte's thought necessarily problematizes any simple dichotomy between the two. On the one hand, Fichte's thought seems to be a form of foundationalism: after all, Fichte is seeking to give an account of the *ground* of all possible experience.[16] On the other hand, Fichte's project seems to be antifoundationalist: traditional foundationalism entails the search for some kind of foundation that is *other* than the doubting self and to which the doubting self may appeal in order to put an end to its doubt.[17] By contrast, the first principle or "foundation" of Fichte's philosophy is nothing other than the questioning, doubting, self-consciously fallible self that knows that no given content can be necessarily determinative for it. Unlike traditional foundationalism, Fichte's system does not provide us with any kind of reference point that is *other* than the self and in relation to which the self might immunize itself (or immunize some privileged set of claims) against doubt. If Fichte's system is to be understood as a form of foundationalism, then it is a foundationalism that forces us through its very enactment to question the very meaning of philosophical foundations.[18]

To ask whether Fichte is a foundationalist or an antifoundationalist is somewhat similar to asking whether Socrates is wise or ignorant. In one respect, it seems that Socrates is truly ignorant: Socrates does not possess knowledge of any determinate content or standard that will allow him to settle any particular question once and for all. But Socratic ignorance does not amount to a complete absence or obliviousness. After all, Socrates is *self-consciously ignorant*, and to that extent he is wise. More pointedly, Socratic wisdom is nothing other than the ongoing activity of self-conscious ignorance. The "content" of Socratic wisdom (self-conscious ignorance) is simply the nonrepresentational awareness that the rational self's thinking can never be fully satisfied or determined by any given content or representation; it is an awareness that the questioning self is always already "beyond" being determined by any thing or representation as it presents itself within experience. In general, it is not possible to understand the meaning of Socratic wisdom without understanding how it implies, and is implied by, Socratic ignorance; by the same token, it is not possible to understand Fichte's foundationalism without understanding how it implies, and is implied by, his antifoundationalism.

The nonfoundational foundation of Fichte's thought is the nonrepresentational, nonsubstantialist activity of the self as self-conscious of its radical fallibility and freedom. By virtue of its radical fallibility and freedom, the self is absolutely unable to explain itself as caused or conditioned by anything that is allegedly independent of it. But this inability of the self to explain itself by reference to something that is allegedly independent of it does not lead to any kind of incipient solipsism. In fact, the activity of the self as we have unpacked it thus far demonstrates that the self (as long as it is a self at all) is necessarily *not* the totality of all that is, and is necessarily finite and limited by what is *other* than itself.

As self-conscious of its radical fallibility and freedom, the self knows that no given content is necessarily determinative for itself, that no given content necessarily imposes itself on the self. However, one "thing" that does "impose" itself on the self is the fact that the self must always *come-to-be* aware of itself as radically fallible and free. The self's coming-to-be as a self-consciously fallible and free self always "happens" to the self, apart from any deliberate or free choosing by the self. The self cannot deliberately and self-consciously choose its own coming-to-be-aware of itself as radically fallible and free (and thus cannot choose to come-to-be the self that it is), since—"prior" to this coming-to-be—the self is "not yet" a self-consciously free self at all. The self-consciously free self is what it is only to the extent that it emerges, or awakens, *out of* a "prior" state of *not* being a self-consciously free self. Since the self was not *always* the radically free and self-conscious self that it is, the self cannot be the totality of all that is, for coming-to-be necessarily implies some otherness. The self-positing self thus cannot be the totality of all that is, and there must be some other to the self, or a not-self (*Nicht-Ich*). With this, we have arrived at the second principle of Fichte's *Grundlage*.[19]

The necessity of the not-self for the self can be explained with reference to the activity of question asking: all question asking presupposes some sense of otherness. As long as any question is not yet answered, there is some other to the self or (what amounts to the same thing) some other to the self's awareness of itself as a self. Insofar as there is some otherness to the self-conscious self, the self is not the totality of all that is, and there must be a not-self. Of course, one might challenge this conclusion by suggesting that the sense of otherness contained in any question does not pertain to any real otherness, but only to an illusory otherness. But even this suggestion confirms the necessity of the not-self for the self. If the otherness implied by the self's questioning referred to an entirely illusory otherness, then the self's being (its activity of being self-aware in a nonrepresentational way) would already be the totality of all that is; but in that case, the self would already know that much, for it would already know everything about everything by virtue of being self-aware. If that were the case, the self could not even *begin* to wonder whether an otherness were real or illusory. It would already know. Thus even the *appearance* of a *possible* otherness (in the form of any type of question) is necessarily a *real* otherness for a self-positing self whose being (as we have seen) consists in the bare activity of self-consciousness.

There must be a not-self, as long as the self is the fallible, questioning self that it is. Thus when Fichte says that the fallible and free self is aware that no given content is necessarily determinative for it, he is not saying that "anything goes" or that there are no limits on the self at all. Just as

Fichte's philosophy of the self-positing self does not amount to solipsism, so too Socrates' relentless question asking does not amount to a sophistry that respects no genuine otherness. We can now also see why Fichte's claims about the self's inability to know any external "thing-in-itself" do not lead us into a bad Cartesian dualism. There is no need to build a "bridge" between the self and the not-self, since the real existence of the not-self is always already entailed by the self's own questioning.

With the second principle of the *Grundlage*, our understanding of the first principle is necessarily transformed. The pure activity of the self-positing self (an activity in which the self's being and the self's awareness fully coincide) is not simply an already achieved givenness from which we make our beginning. As long as there is any otherness *for* the self, there is necessarily also a *difference* between the self's being and the self's awareness; and as long as there is this difference, the pure *Ich = Ich* is both a starting point *and* a yet-to-be-accomplished endpoint. We have already seen that the Fichtean system is self-consciously circular; we now see that the circle is necessarily turning and enriching itself as we go along. We can also see more fully why Fichte's foundationalism must also be an antifoundationalism: the first principle or foundation from which we begin does not remain fixed and does not continue to mean exactly what it meant for us at the beginning. The first principle or foundation develops and shifts in its meaning, even as we continue to rely upon it as our first principle or foundation.

IV. FICHTE'S SOLUTION TO THE PROBLEM OF SELF-MEASURE

Precisely because of the self's radical fallibility, the self cannot claim to know any external thing as it is in itself, unrelated to and unconditioned by the self's own fallible activity. By the same token, the self (as self-conscious of its radical fallibility and freedom) cannot be explained, or accounted for, by reference to anything that is allegedly other than, or independent of, the self. On the other hand, we have also seen that the self *needs* a genuine other (or not-self) in order to be the self-consciously fallible and free self that it is in the first place. It seems, then, that our account of the self and not-self has led us into a contradiction.

The contradiction infiltrates our very definition of the self. The radical fallibility of the self required us to exercise extreme restraint and define the self as nothing other than the pure act of being for self, where the act of self-awareness and the content of the act of self-awareness fully coincide. On the other hand, the radical fallibility of the self also led us to realize that the self could not be a self at all, unless there were also an other *for* the self (a not-self). Because there must also be a not-self for the self, it follows that the self *cannot* be defined as a pure act of being for self, where the act and the content of the act fully coincide.

The entire *Grundlage der gesamten Wissenschaftslehre* can be understood as a series of attempts to eradicate this fundamental contradiction, yet "without doing away with the identity of consciousness."[20] As the *Grundlage* demonstrates, every attempt to eradicate the contradiction ultimately fails. However, the net result is not merely negative: our awareness of the necessity of the failures yields a system of knowledge, a system that implicitly contains a solution to the problem of intellectual self-measure.

As we have seen, the problem of intellectual self-measure arose a genuine problem in the first place because of the self's implicit awareness of its radical fallibility. Furthermore, the self knows that it is fallible only to the extent that it knows that it is not already the totality of all that is, that there is an other to the self in relation to which the self's knowledge might be measured or tested. Thus all measure requires some relation to an other. On the other hand, we have also seen that—precisely because the self is radically fallible—the self cannot claim to know any other as it is in itself, independent of an unconditioned by the self's own fallible activity. Any other for the self is never a genuine other, but only an other as it has been interpreted and understood (and thus perhaps misinterpreted and misunderstood) by the self. Precisely because of this self-conscious fallibility, the self's activity of self-measurement *requires* a genuine other which, in principle, is also *unattainable*.

Fichte's solution to the problem of selfhood and self-measure is to affirm the necessity of the contradiction. To be a self at all is to be always already for oneself or (what amounts to the same thing) self-aware, self-positing, self-intuiting, self-measuring. But the condition of the possibility that the self be the purely self-positing self that it is, is that there be an other (not-self) *for* the self (i.e., an other within the self's awareness). Thus the condition of the possibility that the self be the purely self-positing, self-measuring self that it is, is that it *not* be purely self-positing or self-measuring.

For Fichte, any attempt to escape the contradiction that infiltrates all selfhood and self-measure is an attempt to escape the inevitable finitude of one's own subjectivity. In order to escape the contradiction, one would have to step outside of one's finite subjectivity and somehow achieve a point of view *above* both finite intuiting subject and finite intuited object. One would have to achieve a God's-eye view by which one could see the self-in-itself and the not-self-in-itself as two separate objects, connected by a third, mediating relation. Stated differently, one would have to intuit deliberately and simultaneously *both* the self-in-itself *and* the self-as-it-intuits-the-other. In order to escape the contradiction, one would have to stand above oneself, or behind one's own back.

For Fichte, such simultaneous intuiting (from above or behind) is impossible for a finite self. There simply is no self-in-itself that is intuitable as separate from the self-as-it-intuits-the-other. After all, there could be no self-in-itself, if the self did not *already* intuit an other; and there could be no other *for* the self, if the self were not *already* for itself. Any self that strives to hover above, or get behind, both self and not-self in order to intuit a fixed relation between the two is implicitly trying to intuit something that is external to one's *own* self (here the external thing is the totality constituted by self, not-self, and the relation between them), all the while leaving its *own* selfhood out of the picture. In thus trying to intuit both self and not-self as two independent things in relation to one another, the self inevitably turns these two "things" into a *new*, single object *for itself*. The self and not-self (as two allegedly independent things standing in relation to one another) now collapse into *one* external objectivity for the self that had tried to intuit them as two independent, external things in relation to one another. In short, any alleged God's-eye view (insofar as it is a view of anything determinate at all) can never be a real God's-eye view (a view of self and not-self as two independent things), but has always already become a finite self's view, a view that intuits a single objectivity *other* than itself, all the while leaving its own current activity as a self out of account.

The self-measuring self never knows itself-in-itself apart from knowing itself-in-relation-to-the-other. Because of this, the self-measuring self is always already caught in a contradiction; the self is always already measuring itself by itself *and* in relation to another, *both at once*. However, precisely because of its radical finitude, the self's first impulse is to think that its act of measuring itself is unproblematic and noncontradictory; the self's first impulse is to think that it is measuring itself by reference to a purely external standard, a standard that is independent of and unconditioned by its own activity as a self. This *must* be the self's first impulse, since the self *starts* with a question, a sense that it does not already know everything.

There is indubitably an other for the self, and it is natural for the self to think at first that its act of measuring itself in relation to an other is an act of measuring in relation to a *pure* other.

The finite self can realize only *after the fact* that the other by which it measured itself is not the absolutely independent other that the finite self at first took it to be. This realization *must* come after the fact, since the finite self cannot achieve a simultaneous intuition of both itself and itself-in-relation-to-the other; at first, the self must think that it is simply intuiting a pure other. The finite self realizes only later that the allegedly independent other was actually only an other-for-the-self; it realizes only later that the intuited other was always already an equilibrium of self-and-other together. For Fichte, all subjectivity is finite. This finitude entails some kind of otherness, externality, or difference. If one wants to give this finitude its proper due, then the otherness or difference should not be understood as extended across *space* between two independent things (self and not-self); the otherness extends rather across *time* between two different moments of the same ongoing process (the self as it reconsolidates, or re-collects, itself out of what it first took to be wholly other than itself—but never really was).

With this, Fichte inverts the Lockean account of intellectual self-moderation. According to Locke, the mind *first* discovers its limits, *then* decides to moderate itself according to its knowledge of those limits. Fichte's criticism tells us that, if the Lockean account were correct, then the alleged limits on the self could not be *genuine* limits at all, but only limits as they are understood and interpreted (and thus perhaps misunderstood and misinterpreted) by the self. In other words, such limits would be limits *within* consciousness, and thus not genuine limits at all. For Fichte, any genuine limits on the self cannot simply be the limits that the self finds within empirical consciousness. Genuine limits on the self must be limits that the self has already, unselfconsciously set for itself. In other words, the self does not *first* discover its limits, *then* moderate itself in accordance with them; rather, the self is so radically finite that it is always already self-limiting, and it can discover the limits that it has set for itself only after the fact.

Even after acknowledging the inadequacy of naive realism, the realist philosopher may still seek to hold onto some notion of an object or standard "in itself," unconditioned by all consciousness. The realist philosopher may even offer a sophisticated theory to account for our inability to achieve final and definitive knowledge of any "thing-in-itself." According to the realist, the thing-in-itself always escapes or recedes from our limited grasp, by virtue of its richness or depth or impenetrability. Indeed, experience seems to confirm the realist's account: through experience, we notice that the more we attempt to capture the thing-in-itself, the more the thing-in-itself seems to recede from our grasp. According to the realist, the thing-in-itself has not (yet) been captured by us, but it is at least *logically* possible that one day we may capture it.

For Fichte, the idealist knows that the real reason for the appearance of the thing's noncapturability is not the thing-in-itself, but rather what the self does to the thing in seeking to capture it. As soon as the self has captured anything at all (by bringing it within consciousness), the thing is inevitably no longer a thing-in-itself, but only a thing-for-the-self. There *appears* to be an ever-receding thing-in-itself, not because of anything genuinely independent of the self, but because of the self's own activity. The self *needs* to orient itself toward an other, in order to be a self at all; but as soon as the self knows the other at all, the self has (Midas-like) turned the other into an other-for-the-self; the self has always already destroyed the thing's independent character. The idealist recognizes that the self's own activity is the reason for its perpetual striving. By contrast, the activity of the realist is like the activity of a dog that unselfconsciously chases its own tail, hoping one day to catch it.

In order to be a self at all, the self needs a not-self, a nonempirically given other in relation to which the self is the finite self that it is. This other can never be captured or possessed by the self, as long as the self is a self at all. This points to yet another way in which Fichtean and Socratic philosophy coincide. Socrates acknowledges that all intelligible discourse whatsoever depends upon our awareness of the Forms as the goal or object of thought.[21] And yet Socrates also refuses to give any definitive account of the Forms. For to give a definitive account of the Forms would be to imply that human thought can capture or master that in relation to which it is finite—in which case human thought would not be finite any more. It is precisely because Socrates is so self-conscious of his own limits as a thinker that he refuses to give a definitive account of the Forms. Of course, such refusal *appears* as arrogant to many of Socrates' interlocutors; such refusal *must* appear as arrogant, precisely because the interlocutors *themselves* are arrogant enough to expect that the Forms *can* be captured and made available to human thought once and for all. Like Socrates' self-conscious modesty, Fichte's idealism must *appear* as arrogant to the ones who are truly arrogant.

The Socratic doctrine of Recollection confirms my account of Socratic modesty. Because Socrates is aware of his own ignorance, he realizes that he cannot already have any *actual* knowledge of the Forms. But because of his modesty, Socrates also realizes that he cannot claim that the Forms are *entirely* other than the self. For that claim would implicitly amount to the immodest claim that the self can know an other as it is in itself, entirely unconditioned and independent of the self's own activity. Because Socrates respects the otherness of the other so much, he cannot claim to *know* the other as *wholly* other; hence, the doctrine of Recollection.

The Fichtean self must strive to achieve knowledge of the other as it is in itself, all the while realizing that such knowledge can never be actualized. Of course, the critical reader may ask why the self needs to strive in this way at all. For Fichte, there can be no theoretical answer to that question, but only a *moral* one. The self must strive, because the self cannot be a self without striving, and the self *ought* to be a self. This reasoning is surely circular, but the circularity may appear less vicious if one realizes that the problems of selfhood and self-measure originally arose as problems for the self only because of a moral, and not theoretical, intuition.

The problems arose for the self because the self came to realize that it is radically fallible as a knower, that the self can *never* capture any thing-in-itself once and for all. How did the self come to realize that? Such a realization cannot be based on experience alone, since experience alone yields knowledge only of things as they have been thus far, not of how things *must* be for all time. The self's certainty that it can *never* immunize itself against theoretical error is not a function of the way external things have appeared to the self; it is a function of the self's orientation towards external things. It is not a theoretical certainty, but a moral certainty. Stated more precisely, the self's *theoretical uncertainty* (its awareness that it will *never* be able capture any thing-in-itself once and for all, without the least possibility of correction or revision) is grounded in a *moral certainty*. Along these lines, Socrates argues that we must continue to seek knowledge of the Forms—even when we realize that they may not be capturable by us— because of moral, and not theoretical reasons: ". . . we should be better, braver, and more active men if we believe it right to look for what we don't know than if we believe there is no point in looking. . . ."[22]

The self's striving (in vain) to eradicate the self-contradiction of consciousness arises out of a moral demand; but this striving also happens to serve a theoretical purpose. For in seeking to eradicate the self-contradiction of consciousness, the self (whose activity constitutes the content of the *Grundlage*) unself-consciously generates a series of thought categories (e.g., limitation, quantity, etc.). When all the possible avenues for eradicating the self-contradiction of consciousness have been exhausted, the transcendental deduction of the categories is complete. Fichte can claim that his own deduction of the categories is genuinely *transcendental*, since the content of the deduction emerges solely from the self's ongoing struggle with its self-contradictory, self-conscious fallibility. The self's awareness of its self-contradiction as a self impels it to generate the categories of thought—categories in accordance with which the self had *already* been thinking about the contradiction in the first place. Thus the striving self is engaged in developing a system that, at first, could not appear *as* a system.

Perhaps a final, hermeneutical point is in order. Throughout this essay, I have made reference to Socrates in order to illuminate the meaning of Fichte's thought. The critical reader may reasonably ask whether such continuing reference to Socrates has rendered this a violent interpretation of Fichte's philosophy. I would say that the question erroneously assumes that the figure of Socrates can have meaning apart from our interpretation, and that we can be the philosophical selves that we are apart from the figure of Socrates. First of all, it is clear that there can be no Socrates-in-itself, just as there can be no thing-in-itself. But more importantly, we ourselves cannot be the self-conscious, questioning philosophers that we are without the figure of Socrates. No matter how autonomous and detached we might believe ourselves to be as "critical" philosophers, we are always already philosophizing within a given tradition, namely the tradition engendered by the questioning of Socrates. The real issue for us, then, is not whether we philosophize within that tradition, but whether we do so self-consciously. Indeed, philosophy itself can be understood as the never-ending activity of waking up and becoming self-conscious about what must have always already taken place "behind our own backs" as self-conscious subjects. This essay has been an attempt to contribute to the ongoing awakening.

NOTES

1. John Locke, *An Essay Concerning Human Understanding*, ed. with an introduction by Peter H. Nidditch (Oxford: Clarendon Press, 1975), p. 45.

2. Ibid., p. 46.

3. J. G. Fichte, "Versuch einer neuen Darstellung der *Wissenschaftslehre—Vorerinnerung*," in *J. G. Fichte-Gesamtausgabe der Bayerischen Akademie der Wissenschaften*, eds. Reinhard Lauth, Hans Gliwitzky, and Erich Fuchs (Stuttgart–Bad

Cannstatt: Frommann-Holzboog, 1964ff.), I/4: 185; this English translation is taken from Daniel Breazeale, ed. and trans., *Introductions to the Wissenschaftslehre and Other Writings* (Indianapolis: Hackett, 1994), p. 5.

4. "Versuch einer neuen Darstellung der *Wissenschaftslehre* Vorerinnerung," Fichte, *Gesamtausgabe*, I/4: 185; Breazeale, *Introductions*, pp. 5–6.

5. It is worth emphasizing here that my argument has to do with the modesty of Fichte's system of idealism, and not necessarily with any psychological characteristics that may or may not belong to Fichte as an individual human being.

6. See Plato's *Charmides* (158c) where Socrates asks Charmides whether he possesses the virtue of *sophrosyne*. Charmides cannot answer affirmatively, for such an answer would be immodest. Instead, Charmides remains silent about his self-moderation—a sign that he does, indeed, possess *sophrosyne*.

7. See Plato's *Republic*, 430e–431a.

8. Throughout this essay, the terms mind, intellect, rational self, and knower (and all of their derivatives) are used interchangeably. My use of these terms (and their derivatives) is not intended to suggest any special technical meaning.

9. Of course, while intellectual *self*-measure seems to be impossible, there is nothing to prevent the mind from measuring or testing one representation *within* itself by appealing to some *other* representation within itself as the standard for such measure. However, such an activity would be the testing of one *part* by another *part*, and not the genuine *self*-testing of the self *qua* self.

10. *Fichte: Foundations of Transcendental Philosophy—Wissenschaftslehre* (Novo Methodo), trans. and ed. Daniel Breazeale (Ithaca: Cornell University Press, 1992), p. 89.

11. "Aenesidemus-Rezension," Fichte, *Gesamtausgabe*, I/2: 41; this English translation is taken from Daniel Breazeale, ed. and trans., *Fichte: Early Philosophical Writings* (Ithaca, N.Y.: Cornell University Press, 1988), p. 59.

12. See "Aenesidemus-Rezension," Fichte, *Gesamtausgabe*, I/2: 57; this English translation is taken from Breazeale, *Fichte*, p. 71. For more on the necessarily "dogmatic opposition" between Fichte's idealism and other philosophical viewpoints, see Fichte, *Grundlage der gesammten Wissenschaftslehre* (1794–95), in *Gesamtausgabe*, I/2, 328; *Fichte: The Science of Knowledge*, trans. John Lachs and Peter Heath (New York: Appleton-Century-Crofts, 1920), p. 164.

13. Fichte, *Grundlage der gesammten Wissenschaftslehre*, I/2: 260. This English translation is taken from *Fichte: The Science of Knowledge*, p. 99.

14. Any attempt to account for the self's awareness of itself as radically fallible and free by reference to something other than itself is not only immodest, but also question-begging. For there can be no representational content *for* the self, and thus no explanatory principle or cause known as other to the self, unless the self is "*already*" a self and thus implicitly aware of itself *as* a self.

15. Fichte, *Erste einleitung zu einer neuen Darstellung der Wissenschaftslehre*; *Grundlage der gesammten Wissenschaftslehre*, I/4: 200. This English translation is taken from *Fichte: The Science of Knowledge*, p. 21.

16. See, for example, Fichte, *Erste einleitung zu einer neuen Darstellung der Wissenschaftslehre; Gesamtausgabe,* I/4: 186.

17. The alleged foundation in traditional foundationalism *must* be other than the doubting self qua doubting self; otherwise, it could not put an end to the doubting self's doubt.

18. For more on the question of Fichte's foundationalism and/or antifoundationalism, see the essays by Daniel Breazeale and Tom Rockmore contained in the collection, *Fichte: Historical Contexts/Contemporary Controversies,* eds. Daniel Breazeale and Tom Rockmore (Atlantic Highlands, N.J.: Humanities Press, 1994), chaps. 3, 5.

19. For more on the claim that this "prior" state of inactivity corresponds to the sphere of the not-self, see Fichte: *Foundations of Transcendental Philosophy—Wissenschaftslehre (Novo Methodo),* pp. 121–33.

20. See Fichte, *Grundlage der gesammten Wissenschaftslehre,* I/2: 269; *Fichte: The Science of Knowledge,* p. 107.

21. See Plato's *Parmenides,* 135c.

22. See *Meno,* 86b-c.

5

REFLECTIVE JUDGMENT
AND THE BOUNDARIES OF
FINITE HUMAN KNOWLEDGE
The Path Toward Fichte's
1794–95 Wissenschaftslehre

Arnold Farr

INTRODUCTION

In "Concerning the Concept of the *Wissenschaftslehre* or, of So-called Philosophy," Fichte writes:

> The author remains convinced that no human understanding can advance further than that boundary on which Kant, especially in the *Critique of Judgment*, stood, and which he declared to be the final boundary of finite knowing—but without telling specifically where it lies.[1]

While Fichte does not provide a detailed explanation of the relationship between the development of his philosophy and the development of Kant's thought in the *Critique of Judgment*, it is quite reasonable to believe that the methodological and structural development of Fichte's *Wissenschaftslehre* is, at least in part, the result of a (perhaps brief, but) insightful encounter with the *Critique of Judgment*. One may retort that the notion of reflective judgment does not play a central role in Fichte's philosophy and is rarely mentioned. However, it is not my intention to argue for any explicit use of the notion of reflective judgement in Fichte's philosophy. It is my intention to show that reflective judgment as a principle must be, in some way, taken up by Fichte in his account of consciousness and the limits of human knowledge if he is to advance Kant's philosophy by making it more consistant and systematic.

The 1794–95 presentation of the *Wissenschaftslehre* was an attempt to complete the Kantian project by showing specifically where the boundaries of finite human knowledge lie. Such an attempt must take into consideration what has been taken to be (by Kant anyway) Kant's most successful attempt to demonstrate the unity of theoretical and practical reason, and the limits of finite human knowledge. In this paper I shall examine the role of reflective judgment as a principle wherein the boundaries of finite human knowledge are realized, and I shall argue that this principle is the boundary on which Kant stood, and that it provided Fichte with the insight that gave birth to the *Wissenschaftslehre*. The arguments and methodology of the *Grundlage der gesammten Wissenschaftslehre* will not be explored here. In this paper I simply want to examine Fichte's philosophical development before the publication of the *Grundlage*, and attempt to point out why Kant's position in the *Critique of Judgment* failed to satisfy Fichte and others.

I. THE AWAKENING

Although it was during the summer of 1790 that Fichte was awakened from his "dogmatic slumber" by Kant's practical philosophy (as Kant before had been awakened by Hume) and was working on an explanatory summary of the *Critique of Judgment* by August of that same year, the real turning point in Fichte's thinking occured in 1793. This year marks the publication of *Aenesidemus*. *Aenesidemus* convinced Fichte that, in spite of the efforts of Kant and Reinhold, philosophy was not yet a science. We may say that Fichte was awakened twice. The first awakening in 1790 revealed to Fichte the truth of the Kantian philosophy and its superiority to other philosophical systems. The second awakening in 1793 revealed to Fichte the shortcomings of Kant's philosophy and the need for a new presentation of the Critical Philosophy. Hence, the enthusiastic student of Kant's philosophy would become its critic and the erector of a new system.

In a December 6, 1793, letter to Niethammer, Fichte states that Kant merely pointed to the truth without exhibiting it or proving it. Fichte says of Kant:

> Kant demonstrates that the causal principle is applicable merely to appearances, and nevertheless he assumes that there is a substrate underlying all appearances—an assumption undoubtedly based upon the law of causality. Whoever shows how Kant arrived at this substrate without extending the causal law beyond its limits will have understood Kant.[2]

He continues:

> There is only one original fact of the human mind, a fact which is the
> foundation of philosophy in general, as well as of its two branches, the-
> oretical and practical philosophy. Kant was certainly acquainted with this
> fact, but nowhere did he state it. Whoever, discovers this fact will present
> philosophy as a science.[3]

There are three questions that I will raise with respect to these passages:
First, why does Kant assume that there is a substrate underlying all appearances, and what is this substrate? Second, how does Kant arrive at this substrate without extending the causal law beyond its limits? Third, what is the
original fact of the mind and how does Fichte discover it?

That there is a substrate underlying all appearances is a "necessary maxim
of judgment" which satisfies the speculative (theoretical) use of reason.[4]
Reason in its theoretical use seeks the unconditioned for every conditioned.[5] Reason in its practical use seeks to be unconditioned. Kant writes:

> [We said that] reason, when it considers nature theoretically, has to
> assume the idea that the original basis of nature has unconditioned
> necessity. But when it considers nature practically, it similarly presupposes
> its own causality as unconditioned (as far as nature is concerned), i.e., its
> own freedom, since it is conscious of its [own] moral command.[6]

With respect to reason in its theoretical use it is necessary that we
assume a substrate underlying all appearances in order to know nature as
an organized system. Yet human beings are also required by the laws of
freedom to exercise their freedom in the domain of nature. Kant's task in
the third *Critique* is to show that we must think of nature as if there were
a supersensible substrate underlying nature, which does not conflict with
human freedom, but instead, aids human freedom insofar as nature and
humanity share a common telos (humanity as an end).

In judging nature as a teleological system, the incompatibility between
the laws of nature and the laws of freedom is overcome through the unification of both sets of laws under a common principle. This teleological
principle is a specific form of causality. That is, it is the principle of final
causes rather than efficient causes, because we seek here a form of causality
wherein cause and effect are interchangable.[7] In this respect nature is free
in the same manner that human beings are free; that is, nature is its own
cause and effect. Nature is viewed as a self-organizing whole with respect

to a purpose. The laws of nature are nature's own laws given to nature by itself for the sake of some purpose which nature has determined. We must keep in mind that Kant does not arbitrarily unify or combine the laws of nature and freedom under a common principle, but simply maintains that our scientific investigation of nature demands a unifying principle that is not merely mechanistic. The idea of the purposiveness of nature is not a construct that Kant uses to bridge the apparent gap between the laws of nature and the laws of freedom, but rather, it is a regulative idea that is necessary or useful for the scientific interrogation of nature.

In eighteenth-century science, mechanistic explanations were not always viewed as successful explanations of nature since they merely helped scientists explain organization in nature, but did not help them explain the origin of that organization.[8] John Zammito claims that for Kant mechanical accounts failed to make sense of organic form. To fully explain organic form required the invoking of some nonmechanical cause.[9] Mechanistic theories can explain the way in which phenomena are organized in nature but they cannot explain why and how there is such organization. Mechanism must be subordinated to some nonmechanistic organizing principle whereby we may understand the mechanistic order in nature in terms of nature's capacity for self-organization. That is, we must seek the highest principle whereby the mechanical laws of nature form a complete systematic unity.

Since we are trying to discover the highest organizing principle for the laws of nature, our inquiry has been theoretical; i.e., we are concerned with the possible cognition of an unconditioned, originary principle whereby nature is ordered. However, the principle invoked by Kant is neither cognizable nor simply theoretical, and is thinkable through analogy only. It is in the domain of human moral activity where we encounter an activity that is its own cause (and effect). Human beings act in accordance with final purposes which they themselves determine. It is the nature of reason to subordinate to itself all that is not rational. In "Some Lectures Concerning the Scholar's Vocation" Fichte claims that "[m]an's final end is to subordinate to himself all that is irrational, to master it freely and according to his own laws."[10] By subordinating the irrational to the rational, human beings attempt to fulfill their supreme goal, which is to be in harmony with ourselves. This harmony is possible only if our external world is subordinated to our internal laws. That is, it is only in human beings that we recognize the law of final causality. Here I mean causality in the sense of final causes, freedom, or the power to initiate a series in accordance with a goal. In nature we recognize merely mechanistic causality. Nevertheless, the causality that we recognize in ourselves, in our humanity, demands that we exercise causality in the realm of nature. That is, the demand that we freely determine ourselves is the demand that we determine nature insofar as any act of self-determination requires a realm wherein such an act may occur. I can only determine myself by acting upon nature, which is the domain for my moral activity. Nature, however, is cognized according to certain rules of the understanding. The way in which nature must be cognized according to these rules problematizes the demand that we exercise causality in the world to the extent that if nature is determined by its own laws it cannot be determined or modified by the laws of human freedom.

According to Kant, nature can be cognized only as mechanism. Kant claims that "without mechanism we cannot gain insight into the nature of things."[11] However, the view of nature as mechanism does not satisfy the demands of reason. Kant states that "[r]eason is a power of principles, and its ultimate demand [for principles] aims at the unconditioned."[12] Nature as mere mechanism does not satisfy the demand of reason because it is based on the notion of a "causal connection that does not necessarily presuppose the understanding of a cause."[13]

Since we perceive only contigent causal connections when we view nature as mere mechanism, we must assume a second view of nature wherein we can at least think of a causality that is characterized by more than contigent causal connections. We seek a causal principle that does not lie in nature, but is the basis for the possibility of nature. That is, we seek a supersensible substrate of nature that unifies all laws of nature into an organized systematic whole. The idea of this supersensible substrate or originary principle is provided by the principle of reflective judgment.

Through reflective judgment we must think of nature as purposive, we must view natural products not as merely produced by physical (mechanical/efficient) causes, but as the products of final causes. However, we do not know nature as purposive but must simply think of it as being so. Reflective judgment is a subjective principle that tells us nothing about nature and does not affect nature but only affects the human subject. It reveals nothing to us about the laws of nature, but it does reveal something to us about the laws of our own thinking (which are explained in detail by Fichte). It reveals that the demands of reason are not only given by the subject, but can be met only within the subject. In other words, nature does not necessarily conform to the demands of reason, although it is our vocation (*Bestimmung*) to bring nature in line with reason for theoretical and practical purposes. Insofar as the human subject is incapable of determining nature, it must instead determine its thinking about nature. We give ourselves a principle whereby nature is thought to be in harmony with our concept of purpose.

What have we discovered in the foregoing? The scientific investigation of nature requires that we think of nature in two ways. First as a manifold of appearances to which the causal principle is applicable. Here, we view nature merely as mechanism. In the mechanistic theories of nature the causal principle that is applicable to nature is an objective principle for determinative judgment[14] supplied by the understanding. According to this principle nature acts causally but not intentionally. To be sure, this causal principle is applicable to appearances only because its concepts are given by the understanding that cannot go beyond experience. Nevertheless, insofar as reason seeks the unconditioned (or the originary organizing principle for all natural organisms), we inevitably strive to go beyond experience. For this reason Kant claims that we must assume that there is a substrate wherein all of the laws of nature are unified according to a "subjective principle for merely reflective judgment and hence a maxim imposed on it by reason."[15] Therefore, the second way in which we must think of nature is as an organized and organizing system which organizes itself in accordance with a purpose.

Consequently, the highest knowledge of nature available to us is merely a regulative way of thinking about nature based on a subjective principle. Hence, we are unable to get beyond our own subjectivity in our drive to establish our knowledge of nature as an organized and organizing system. Where then do the boundaries of finite human knowlwdge lie? They lie within the subject. Kant may very well have known this but failed to offer an adequate proof. Kant put subjectivity into nature while Fichte put objectivity into the subject, thereby, completeing the Copernican revolution.

The subjective principle of the third *Critique* is the starting point for the *Wissenschaftslehre*. Fichte tells Reinhold in an April 28, 1795, letter: "All I had to do was combine Kant's discovery, which obviously points in the direction of subjectivity, with your own."[16] This direction of subjectivity toward which Kant's discovery points combined with Reinhold's discovery (that philosophy as a system must be based on a first principle) led Fichte to a discovery of the "original fact of the mind," which will be discussed shortly.

In a July 2, 1795, letter to Reinhold, Fichte claims that "Kant seeks to discover the basis for the unity of the manifold in the not-I."[17] While Kant's discovery points toward subjectivity, it fails to treat adequately the nature of subjectivity. Kant's mistake was in his attempt to ascribe to nature (the not-I) a supersensible substrate through the principle of reflective judgment. However, Kant's discovery points toward subjectivity only insofar as this supersensible substrate ascribed to nature is provided by the I (the human subject).[18] Hence, if there is in experience a supersensible substrate, then it is solely within the I. What has to be explained then is how the I, as the only supersensible substrate of experience, accounts for its feeling of limitation; i.e., how do we explain the affection of the I by nature?

In a letter dated July 4, 1797, to Reinhold, Fichte claimed that Kant never posed the question of external sensation.[19] Although this claim was made after the first presentation of the *Wissenschaftslehre*, it was made in retrospect. In this letter Fichte is addressing Reinhold's criticisms of the *Wissenschaftslehre*. Fichte was simply trying to make clear what it was that he attempted to accomplish in the *Wissenschaftslehre* and how his project was the same in spirit as Kant's but was a more thorough working out of the problems raised or alluded to by Kant. Entailed in this problem of external sensation is the problem of consciousness and the problem of the nature of subjectivity and objectivity. Indeed, all three are one and the same problem. Further, if we are to understand how theoretical and practical reason form a unity, the abovementioned problems must be explained.

The failure of Kant's presentation of the Critical Philosophy lies in the fact that Kant failed to erect his system upon a first principle of which one could be certain, and from which one could derive all other propositions of a system. Although the standpoint occupied in the *Critique of Judgment* points toward the boundaries of finite human knowing, knowledge of where these boundaries lie specifically requires an absolutely unconditioned first principle of human knowledge. According to Fichte, the truth of any proposition lies in the relation between that proposition and all other propositions of that system. All propositions of a system are held together by a common source or ground (a first principle). Since all knowledge requires a unity of consciousness, the task of the *Wissenschaftslehre* is to discover the basis for all consciousness. We discover that all consciousness is based on self-consciousness, and self-consciousness is made possible only through the feeling of limitation that the I encounters through its activity. Fichte explains consciousness of apparent external sensations by explaining the actual activity of the finite human I. It is by means of this explanation of the activity of the finite human I that we determine precisely where the boundaries of finite human knowledge lie.

Fichte's strategy is first to abstract from all empirical consciousness in order to discover the basis for all consciousness. This basis for all consciousness cannot be a fact of consciousness because facts of consciousness appear among empirical states of consciousness. We discover that the basis of all consciousness is not merely a fact, but rather, an act (this act is the original fact[20] of the mind, which is the foundation of all philosophy, and with which Kant was acquainted but did not state). The notion that all consciousness is based on an act is Fichte's most original contribution to philosophy, and marks his advance beyond Kant's philosophy. It is with this notion that Fichte develops a theory of subjectivity that explains our feeling of objective necessity. It is difficult to say when Fichte first conceived the idea that the basis of all consciousness is an act, but it is apparent that such an idea came to full fruition in 1793 while Fichte worked on his review of *Aenesidemus*.

II. TOWARD A THEORY OF SUBJECTIVITY (OBJECTIVITY): THE *AENESIDEMUS* REVIEW

G. E. Schulze's *Aenesidemus* was instrumental in making the problems with Kant's Critical Philosophy transparent to Fichte. The preponderance of Schulze's criticisms of the Critical Philosophy was directed at Reinhold's formulation of it, which Schulze took to be its strongets and most systematic formulation. It was Schulze's criticisms of the Critical Philosophy that made Fichte more aware of its structural and methodological problems.

Fichte agreed with Reinhold that the Critical Philosophy should be founded on a first principle and he also agreed that the first principle should be the principle of consciousness. However, it is the nature of this Principle of Consciousness that is the point of contention. It was Reinhold's formulation of the Principle of Consciousness that was attacked by Schulze. Schulze thought that his criticisms of Reinhold's and Kant's presentations of the Critical Philosophy would destroy the Critical Philosophy in general. We shall see that this was not the case.

Fichte agreed with many of Schulze's criticisms but thought that the Critical Philosophy could be salvaged if made more methodologically sound. In part 1 of the *Grundlage* Fichte employs a method that overcomes Schulze's attack. This method will not be dealt with here. Instead, I simply want to examine some of the developments in Fichte's "Review of *Aenesidemus*," which contains Fichte's first principle in an embryonic stage.

Schulze's first criticism of Reinhold's philosophy was that Reinhold's first principle was not a first principle at all since it is subordinate to the principle of contradiction.[21] Fichte claims that Reinhold acknowledges this point but claims that the Principle of Consciousness is not subordinate to the principle of contradiction in the way that "a proposition is subordinate to a first principle which determines it, but rather as subordinate to a law which it may not contradict."[22] That is, the principle of contradiction has no material validity but only formal validity.[23]

According to Fichte, Schulze's second criticism of the Principle of Consciousness as the first principle of all philosophy is that it is not determined completely by itself.[24] Fichte agrees with Schulze on this point. Reinhold's Principle of Consciousness asserts that "in consciousness, the representation is distinguished from and related to, the subject and object by the subject."[25] The problem with this formulation of the principle is that the concepts of distinguishing and relating are ambiguous. These concepts entail several possible meanings. The Principle of Consciousness must be determined by something outside of itself if these concepts are to be defined unambiguously. Fichte asks:

> But what if it is precisely the indeterminacy and indeterminability of these concepts which point to a higher principle (which remains to be discovered) and to the material validity of the principle of identity and opposition? And what if the concepts of distinguishing and relating can only be determined by means of the concept of identity and its opposite?[26]

These questions mark a crucial insight for Fichte, which would lead him to the discovery of his own first principle of all philosophy and would circumvent Schulze's attack on the Critical Philosophy. In the *Grundlage* Fichte replaced the concepts of distinguishing and relating with the principle of identity ($A = A$) and the principle of opposition ($A = {\sim}A$). When given content, the principle of identity is expressed as $I = I$ and the principle of opposition is expressed as $I = {\sim}I$. Fichte's shows that the principles of identity and opposition are necessary conditions for any consciousness whatsoever. However, this will be discussed in more detail later. For the moment we must examine Schulze's third criticism of Reinhold's principle of consciousness, and then examine the way in which Fichte responds to these criticisms en route to his formulation of the first principle of philosophy.

Schulze's third criticism of Reinhold's Principle of Consciousness is that it is not universally valid. That is, there are states of consciousness to which Reinhold's principle does not apply. Fichte writes:

Aenesidemus (Schulze) adduces several empirical manifestations of consciousness which, in his view, lack the three elements which are supposedly required for all consciousness [viz., a representation, a representor, and that which is represented].[27]

He continues:

> If no consciousness is conceivable apart from these elements, then they are of course all included in the concept of consciousness, and of course the proposition which asserts this is, with respect to its logical validity as a proposition based upon reflection, an analytic proposition. Yet since it involves distingushing and relating, this very action of representing, the act of consciousness itself, is obviously a synthesis and the foundation of all other possible synthesis.[28]

Schulze's third criticism of the principle of consciousness made Fichte realize that our knowledge of the three elements of consciousness (as delineated by Schulze) is possible only through an act of reflection. It is through reflection that the three elements of consciousness are distinguished. Further, the proposition which designates these three elements as belonging to consciousness is an analytic proposition.

In the *Grundlage* Fichte defines analytical thinking as "the act of seeking in things equated the respect in which they are opposed."[29] This very act, however, requires an initial synthesis wherein the opposed or distinguished elements are unified or alike. But even the highest synthesis presupposes an originary opposition. But, how do we arrive at this highest principle? Fichte claims that the Principle of Consciousness is an abstract principle.[30] Reinhold would deny this. However, the Principle of Consciousness only explains the representation of empirical objects. Fichte states that:

> If everything that can be discovered in the mind is an act of representing, and if every act of representing is undeniably an empirical determination of the mind, then the very act of representing, along with all of its conditions, is given to consciousness only through the representation of representing.[31]

This passage expresses Fichte's agreement with Schulze's third criticism of Reinhold's Principle of Consciousness. That is, the Principle of Consciousness fails to account for all manifestations of consciousness.

It will be Fichte's task in the *Grundlage* to discover the highest form of consciousness, and from there, deduce all other manifestations of consciousness (even consciousness of nature). This procedure requires an abstraction from the spacial and temporal mode of empirical representation. In the representation of representing, abstraction from all empirical determinations of an object is necessary.[32] Hence, Fichte's first principle is the highest synthesis, which is arrived at by abstraction from all empirical content of consciousness. It will eventually be shown that consciousness requires empirical content, but that the ground of consciousness must express more than empirical facts; it must also express an act whereby empirical facts are known or produced for consciousness.

The mistake that Reinhold made was to assume that the first principle of all philosophy must be a fact.[33] Fichte claims that Schulze's objections to the Principle of Consciousness is warranted only if it is an objection to the Principle of Consciousness as the first principle of all philosophy and as a mere fact. The Principle of Consciousness in itself is immune to Schulze's criticisms.[34] However, such a principle must be presented as expressing an act, not as expressing a mere fact.

That the first principle of all philosophy should express an act is necessary if the first principle of philosophy is going to account for all manifestations of consciousness. Reinhold's Principle of Consciousness fails to account for self-consciousness. Self-consciousness is the representation of representing, it is a form of consciousness that is not representational, but is the ground for representational consciousness. Fichte writes:

> The Principle of Consciousness, which is placed at the summit of all philosophy, is based on empirical self-observation and certainly expresses an abstraction.[35]

That is, all consciousness is based on self-consciousness. The above passage points to another agreement between Fichte and Schulze. Schulze claimed that Reinhold's "explanation is narrower than what it is suppose to explain."[36] According to Reinhold, a representation is that which is distinguished from subject and object, and yet is related to both.[37] Consciousness is based on this act of distinguishing and relating, but as we have seen, it is this very act that must be explained. Fichte states that:

> Both distinguishing and relating can become objects of representation, and really are such within the context of the Elementary Philosophy. However, they are not representations to begin with, but only the ways in which the mind necessarily must be thought to act if it is to produce a representa-

tion. But of course it undeniably follows from this that representation is not the highest concept for every conceivable act of the mind.[38]

In the above passage Fichte continues to assert that Reinhold's first principle is inadequate as a first principle of all philosophy. However, he continues to share a fundamental agreement with Reinhold to the extent that he rejects Schulze's view that distinguishing and relating (which are necessary for representation) are themselves acts of representing.[39] Fichte's response to Schulze seems rather confusing at first. Nevertheless, it adumbrates Fichte's key insight in the development of his first principle. What does Fichte mean by the claim that distinguishing and relating "are not representations to begin with but only the way in which the mind necessarily must be thought to act if it is to produce representation?"[40] This statement implies that distinguishing and relating may become representations through some higher act of the mind. How, then, is it possible that representations are produced by the act of distinguishing and relating? It can only mean that there is an act of the mind that is higher than representation.

Since in consciousness the representation is distinguished from and related to a subject and an object one may ask, What is immediate in consciousness? For Reinhold, representation is immediate while "subject and object exist in consciousness only by means of the relation of the representation to them."[41] For Schulze, "subject and object exist immediately" and representations are produced mediately in consciousness. Fichte writes:

> The subject and object do indeed have to be thought of as preceding representations, but not in consciousness qua empirical mental state, which is all Reinhold was speaking of.[42]

Fichte's claim is simply, with respect to empirical mental states, that is, consciousness of empirical objects, representation precedes our consciousness of the subject and object. However, this very representation presupposes or is made possible by some nonempirical, nonrepresentational act or condition. Here Fichte formulates his notion of the absolute I and the not-I, which receives full formulation in the *Grundlage*.

In the *Grundlage* Fichte develops a theory that shows how consciousness of subject and object is produced. Hence, Fichte's theory of subjectivity is equally a theory of objectivity.[43] As we have seen, for Fichte, Kant's Critical Philosophy never adequately explained external sensations, and although Kant rightly pointed toward subjectivity, he never explained it. It is only through an arrant explanation of subjectivity that we can explain our feeling of objective necessity. The very concept of nature or an external world is based on this feeling of necessity. Fichte (unlike Kant) does not evoke a subjective principle to unify the laws of freedom and the laws of nature; rather, he tries to explain the feeling of necessity that is found within the subject. It is this feeling of necessity that leads us to posit an external world. Further, Fichte shows how freedom is not possible without limitation.

This leads us to a fundamental problem with the Critical Philosophy that also haunts Reinhold's Elementary Philosophy. The task of the Critical Philosophy is to explain the necessary conditions for any possible experience. The reasons for Kant's alleged failure are too numerous and complex to be dealt with in detail in the space of this paper. However, we have seen that one of the most important problems with the Critical Philosophy that Reinhold and Fichte attempted to remedy was its lack of systematic form. The first step in remedying this problem was to establish a first principle whereupon the Critical Philosophy could be re-erected. Reinhold's Principle of Consciousness was such a first principle. However, this Principle of Consciousness merely shows that consciousness is based on the act of distinguishing and relating. Reinhold does not explain the necessary conditions for this act. It must be explained by something even more fundamental than consciousness, (i.e., something that does not occur among the empirical states of consciousness).[44] There must be some principle that explains the empirical states of consciousness, or a form of consciousness that is not representational but is the ground for representational consciousness. This principle will also explain external sensation. Hence, Reinhold's Principle of Consciousness must be based on an even higher principle.[45]

Fichte's new principle advances beyond Kant's and Reinhold's philosophies because it explains external sensations, our feeling of objective necessity, and representational consciousness. This leads us back to the problem of reflective judgment. Reflective judgment is merely a subjective principle whereby we ascribe subjectivity to the external world in order to assert a unity of two conflicting feelings that we find within us. This principle never demonstrates the unity of the laws of nature and the laws of freedom. It merely asserts that we treat one as if it were the other. The merit of Fichte's philosophy is that it explains how the very concept of nature and the feeling of necessity that accompanies it is produced. Hence, Fichte achieves what Kant did not quite achieve; that is, he explains the necessary conditions for any possible experience (particularly a unified experience).

It is to Kant's credit that he did indeed lay out all of the necessary elements that constitute experience, but never adequately explained how these elements work together to constitute a unified experience. The principle of reflective judgment represents Kant's highest ascent in his attempt to show the unity of experience. Anyone who attempts to overcome the shortcomings of Kant's philosophy must determine the validity and usefulness of this principle. That is, one must explain why the principle of reflective judgment fails as a unifying principle. Fichte merely asserts that the basic idea behind the principle of reflective judgment is correct; that is, Kant's attempt to explain experience and the unity of consciousness through a subjective principle is right. However, the ultimate shortcoming of the principle of reflective judgment is that it does not explain external sensation or our feeling of necessity.

Fichte recognizes that empirical consciousness is not completely unified. It is at this level of consciousness that we experience a feeling of necessity that comes into conflict with our feeling of freedom. Fichte's strategy is to abstract from all empirical consciousness (representational consciousness). There must be some nonempirical ground for empirical consciousness. Kant and Reinhold both failed to explain empirical consciousness because they failed to properly abstract from it. That is, through abstraction one discovers how empirical consciousness is produced.

For Fichte, consciousness must be explained in terms of self-consciousness. We must begin with self-consciousness because it is what is most immediate and certain. All other consciousness is mediated through self-consciousness. What happens then, to Kant's principle of reflective judgment? In Fichte's philosophy reflective judgment is *aufgehoben*, it is annuled and maintained. It is annuled insofar as for Fichte reflective judgment is no longer required as a way of thinking about nature; that is, we are not required to ascribe a causal principle or intentionality to nature. It is maintained to the extent that we still posses a subjective principle of purposiveness. However, this principle is not ascribed to nature but is recognized as belonging only to the human subject. Cognition of nature as that which is in opposition to human freedom can be derived only from an examination of subjectivity itself. Therefore, Fichte's question is, how are we conscious of nature at all?

According to Fichte, consciousness of the external world—i.e., consciousness of nature—is due to a limitation that the I finds within itself. It is because of this feeling of limitation that the I posits something outside itself as the cause of this limitation. Here is a crucial difference between

Kant and Fichte. For Kant we overcome the conflict between the laws of nature (the source of limitation) and the laws of freedom (the demand that we overcome limitation) by thinking of nature as purposeful. Through reflective judgment we impose the concepts of freedom onto nature. We ascribe a human characteristic to nature. For Fichte, the source of limitation is taken into the I; that is, it is explained as something that is internal to the I rather than external. The conflict between the concept of freedom and the concept of nature is not to be explained as simply a conflict between the I (concept of freedom) and something external to the I (concept of nature), but rather, as a conflict within the I itself. The very being of the I is not possible apart from this conflict. How does this conflict occur within the I, or what are the conditions for self-consciousness or any consciousness whatsoever?

The conflict that we have been talking about is nothing more than the division that results from the bidirectional activity of the I. Any explanation of consciousness must begin with an explanation of this bidirectional activity because consciousness itself is constituted by this activity. We may approach the problem by asking, What are the necessary conditions for consciousness, i.e., what are the necessary conditions for any finite human I?

III. THE I AS A DIALECTICAL SELF-FORMATIVE PROCESS

We have arrived at what I consider to be one of the most salient features of the Fichtean philosophy, that is, human consciousness is a dialectical self-formative process whereby *das Ich* is formed. While Kant's solution to the problem of the unity of theoretical and practical reason is based on a subjective principle, Fichte begins with the problem of subjectivity itself. Kant's subjective principle does not account for the I as both subject and object. Subjectivity is only one side of the I. We must seek a deeper unitary structure whereby subjectivity itself is made possible, or at least whereby self-knowledge by the subject is possible. Ultimately, subjectivity is formed in relation to (indeed, in opposition to) objectivity. That which is objective is not merely that which lies outside of the I, it is also part of the I.[46]

The conflict between the concepts of freedom and the concepts of nature is not adequately treated by Kant insofar as he does not adequately explain how the I is limited. Whereas for Kant the human moral agent is autonomous and encounters a limit to his or her activity in nature, for

Fichte limitation occurs within the I itself. The opposition is no longer merely between the I and something external to the I, but, the opposition occurs within the I itself insofar as the I is a unity of freedom and nature. The I posits itself as free and limited at once. Indeed, freedom is nothing more than self-limitation. Here, the unitary structure that we are looking for begins to unfold. In the act of self-positing an opposition is posited. This opposition is an act which is the original fact of consciousness. The I is conscious of itself as that which determines and that which is determined, as producer and product. This apparent contradiction is the essential nature of the I. The finite human I is constituted by two opposing activitites which are at work in the I.

This dialectical self-formative process can be explained only within the I itself. Further, the explanantion of this process also explains the I's feeling of freedom and limitation. Everything must be explained in terms of the I's dialectical self-formative process because before we can explain "consciousness of" (representational consciousness) consciousness in general must be explained, and it must be explained in terms of its own rules. That is, the I can never examine an object of consciousness through any means other than an examination of its own self-consciousness. Therefore, the I is not permitted to ascribe causality to anything other than itself. It is only through the I's activity that the external world is experienced. Insofar as the I discovers itself to be the ground of all experience, it discovers itself to be the ground of all laws that govern experience, and also the origin of any purpose. What then, is the status of human knowledge? Human knowledge must begin and end with self-knowledge. The boundary of finite human knowledge is the activity of the I. That is, the I can have no knowledge of nature except through its own activity. Hence, the boundaries of finite human knowledge are ideed subjective as Kant pointed out. However, Kant never showed how at the base of subjectivity lie two opposing feelings (the feeling of freedom and the feeling of necessity) whereby subjectivity and objectivity are constituted. It is with this bold discovery that the path toward the 1794–95 *Wissenschaftslehre* begins.

NOTES

1. J. G. Fichte, *Early Philosophical Writings*, trans. Daniel Breazeale. (Ithaca: Cornell University Press, 1988), p. 95.
2. Ibid., p. 369.
3. Ibid.

4. Immanuel Kant, *Critique of Judgment*, trans. Werner S. Pluhar (Indianapolis: Hackett, 1987), p. 282.

5. Reason, in its quest for a unified sustem of knowledge gives itself two demands which are necessary, yet in conflict with each other. At B644 in the *Critique of Pure Reason* Kant says that: "[t]he one calls upon us to seek something necessary as a condition of all that is given as existent, that is, to stop nowhere until we have arrived at an explanation which is completely a priori; the other forbids us ever to hope for this completion, that is, it forbids us to treat anything empirical as unconditioned and to exempt ourselves thereby from the toil of its further derivation." This passage may be translated into the language of the third *Critique* as follows: Reason seeks the original organizing principle of all organisms. This original organizing principle must not be anything empirical, but must lie beyond all empirical nature. However, we cannot know any causality beyond empirical causality, that is, our knowledge is restricted to a mechanistic view of nature. Nevertheless, we must still seek the underlying principle of all mechanistic causality.

6. Kant, *Critique of Judgment*, p. 286.

7. In section 65 of the *Critique of Judgment*, Kant claims that an efficient cause is the kind of causal connection our mere understanding thinks as constituting a descending series of causes and effects. An effect presupposes something else as its cause and cannot in turn be the cause of the other thing. Mechanism can account only for this kind of causal connection. However, Kant also claims that "we can conceive of a causal connection in terms of a concept of reason (the concept of purpose). Such a connection considered as a series, would carry with it dependence both as it ascends and as it descends" (p. 251). This kind of causal connection is that of final causes. It is only through the idea of a final cause that we can think of nature as purposive, i.e., as a self-organizing system.

8. Peter McLaughlin, *Kant's Critique of Biological Explanation: Antinomy and Teleology* (Lewiston, N.Y./Queenston, Ontario: Edwin Mellen Press, 1990), p. 9.

9. John Zammito, *The Genesis of Kant's Critique of Judgment* (Chicago: University of Chicago Press, 1992), p. 215.

10. Fichte, *Early Philosophical Writings*, p. 152.

11. Kant, *Critique of Judgment*, p. 295.

12. Ibid., p. 283.

13. Ibid., p. 290.

14. Determinative judgment is the power to subsume the particular under the universal. The universal under which the particular is subsumed is given by the understanding. However, the principles of the understanding fail to represent the highest possible unity to the extent that their application is restricted to experience. Reason demands knowledge of the laws of nature as a unified system, which is not given in experience.

15. Kant, *Critique of Judgment*, p. 280.

16. Fichte, *Early Philosophical Writings*, p. 390.

17. Ibid., p. 399. In this letter Fichte discloses to Reinhold that his Principle

of Consciousness merely demonstrates the unity of speculative (theoretical) reason, while the *Wissenschaftslehre* demonstrates the unity of all three of Kant's *Critiques*. Kant also failed to unify all three *Critiques* because his solution is to a great extent theoretical. That is, since the basis of unity is attributed to nature (the not-I) it appears that the I may be determined by the not-I.

18. Kant jumps the gun in his attempt to ascribe to nature a supersensible substrate to the extent that our awareness of any supersensible substrate and our awareness of nature begin with our own activity, which Kant never explains.

19. Ibid., p. 420.

20. Fichte expresses this act that is also a fact by the term *Tathandlung*, which combines the words *Handlung* (action) and *Tatsache* (fact).

21. Fichte, *Early Philosophical Writings*, p. 61.

22. Ibid.

23. In section 1 of the *Grundlage der gesammten Wissenschaftslehre* (Berlin: Walter de Gruyter, 1971), Fichte shows how the principle of contradicition is present in consciousness as a principle that must not be contradicted, but this principle is simply a form of judgment that is produced by a consciousness that has contradiction at its origin. That is, the principle of contradiction may be easily applied to a proposition such as A = A, but it is the finite human I that makes such a judgment. Further, when a formal proposition such as A = A is given content or material validity, such as is the case is in the proposition I = I or I am I, the absolute identity that is implied by the principle of contradiction is split into subject/object poles that embody a contradiction that is necessary if there is to be any consciousness at all. The I as object is passive, while the I as subject is active. The coterminus existence of passivity and activity in the same entity signifies a contradiction in the very constitution of that entity. Consciousness requires the coterminus existence of activity and passivity in the same entity.

24. Ibid., p. 62.

25. Frederick Beiser, *The Fate of Reason: German Philosophy from Kant to Fichte* (Cambridge, Mass.: Harvard University Press, 1987), pp. 252–53.

26. Fichte, *Early Philosophical Writings*, p. 62.

27. Ibid.

28. Ibid., pp. 62–63.

29. J. G. Fichte, *Science of Knowledge*, trans. Peter Heath and John Lachs (Cambridge: Cambridge University Press, 1982), p. 111.

30. Fichte, *Early Philosophical Writings*, p. 63.

31. Ibid.

32. Ibid.

33. Ibid., p. 64.

34. Ibid., p. 65.

35. Ibid., p. 63.

36. Ibid., p. 64.

37. Ibid.

38. Ibid., p. 65.

39. Ibid., p. 64.

40. Ibid., p. 65.

41. Ibid.

42. Ibid.

43. Frederick Neuhouser's book *Fichte's Theory of Subjectivity* (Cambridge: Cambridge University Press, 1990) is misleadingly named insofar as Fichte's is simultaneously a theory of objectivity. That is, it is impossible to explain subjectivity without explaining objectivity, and vice versa. This insight marks Fichte's original contribution to the history of philosophy and his triumphant response to the problem of reflective judgment as a subjective principle. Kant's principle of reflective judgment does not give an adequate account of objective consciousness.

44. Fichte, *Science of Knowledge*, p. 93.

45. Fichte, *Early Philosophical Writings*, p. 64.

46. The laws of nature that limit the I are also necessary for the very existence of the I to the extent that the I is an embodied I. The body is the I's most immediated limitation. However, it is only through the body that the will of the I can be expressed. The body is required for moral activity. The role that the body plays in Fichte's philosophy is not clearly articulated until toward the end of the Jena period in the *Wissenschaftslehre nova methodo*.

IMAGINATION AND TIME
IN FICHTE'S *GRUNDLAGE*

C. Jeffery Kinlaw

There is little doubt that the productive imagination is a fundamental element in Fichte's *Wissenschaftslehre*. The imagination, as Fichte himself insists, is that "wondrous capacity . . . without which nothing in the human spirit can be explained—and on which the entire mechanism of the human spirit can easily ground itself."[1] In fact, as Fichte stated in a public lecture at Jena, the imagination is a synonym for spirit (*Geist*), meaning, as is evident within the *Grundlage*, that the imagination is the movement of thought itself.[2] It does not simply apply the categories of thought as Kant had maintained. Rather, it produces them, for the categories themselves represent modes or types of synthesis that are effected by the imagination.[3] Even though the imagination is derived *as an independent activity* in the course of the *Wissenschaftslehre*, it is clearly presupposed from the beginning, and presupposed as what executes syntheses.[4] Not only is the imagination the necessary condition for the possibility of experience and consciousness, it is likewise the necessary condition for life itself, the sole stabilizing element that gives integrity and coherence to the self. In this sense Fichte's view of the imagination represents a radicalization of the Kantian view, both with respect to its cognitive role as a mediator between otherwise heterogenous concepts and sense data in the First Critique, and its interpretative role of envisioning nature's systematic orderliness in the Third Critique. These two functions, the scientific and the aesthetic, Fichte tends to collapse, as Rudolf Makkreel has indicated.[5] But there is another distinction, one that parallels Makkreel's, that Fichte also collapses, what I will call the scientific and the hermeneutical (the understanding or interpretation of the self as a self). In both cases, to emphasize again, the imagination is the necessary condition for experience and for the stability and integrity of the self.

The imagination's task, which becomes the task for the *Wissenschaftslehre* itself, is to think opposites, that is, to think them as unified in their very opposition. It is the I's inability to think opposition that leads the imagination, as it wavers and oscillates between opposites, or in John Sallis's words encircles them,[6] to fashion an image, a substrate that makes possible the unity of opposites, I and not-I, subject and object, as well as their very positing as opposites.[7] This substrate, though fleeting and unstable, is time itself. Time thus makes possible the relation between subject and object, indeed the very positing of subject and object, that is intrinsic to experience, as well as the relational unity of the I as infinite and finite. In both cases it is the imagination which provides a temporal substrate, a momentary fixation, as a state of the I in sensory experience and in its own self-understanding as a self.

In this paper I intend to argue that the hermeneutical and scientific functions of the imagination are not only interconnected, but that the hermeneutical role, the effort to provide for proper self-determination, to maintain the coherence of the self, is more basic than the imagination's theoretical or scientific role. This claim by itself may be uncontroversial, for this is precisely how one might read the practical part of the *Wissenschaftslehre*. After all, Fichte shows, by the category of striving, that practical reason becomes the basis for theoretical reason and thereby provides the unity for reason that Kant envisioned but never achieved. But I will maintain that the priority for which I will argue is already adumbrated in the theoretical *Wissenschaftslehre*. According to my interpretation, then, the *Wissenschaftslehre* cannot be properly read merely as an epistemological project. If Fichte is to derive the Kantian categories—necessary conditions for the cognition of objects in general—from the self's own attempt to think simultaneously its self-positing and its non-self-positing or positing of an other, then the deduction of the categories is tied to the coherence of the self, namely, the self's identity as an I.[8] I will focus on the way time functions as the *Urbild* that gives stability both to experience and to self-understanding, but, equally important, on the relation of time and imagination as a "beneficent illusion" (*wohltätiges Täuschung*). The illusory nature of the imagination's products will show not only that the scientific imagination serves the hermeneutical, but what is ultimately real for Fichte, and what defines the self as *Geist* is the imagination itself. In a sense, as I will attempt to show, time is a product of the I's own frustration. It provides a momentary resting place for the I on its way to becoming what it means to be an

I as well as a horizon or framework for that self-determination. It is time that gives provisional meaning and coherence to the self-projection and self-reflection that defines the life of the I. My argument will proceed as follows. In the first section I will explicate the role of time and the imagination in the construction of cognitive experience, that is, the scientific function of the imagination in producing our experience of the external world. In section 2, I will show how imagination and time provides the basis for the coherence and stability of the self. Though these functions are interrelated within the *Wissenschaftslehre*, they are separated in my analysis for the purpose of explication. In a concluding section, I will indicate, by reference to the illusory nature of the imagination, why its hermeneutical function, while intertwined with its scientific or epistemological function, is more basic.

I

The task of the imagination is to think opposites and to think them as simultaneously opposed and united. But it can do so only by oscillating between them, and in its very oscillation, its own *Schweben*, imagination fashions an image (*Bild*) that momentarily reconciles mutual opposition. The image or intuition is initially inchoate and amorphous, since it is largely still indeterminate. The imagination's continued oscillation between the opposites it unites in the intuition clarifies the intuition and makes it more precise and determinate. In this way, one not only sees an unspecified object, but one begins to assign certain properties (and exclude others) to the object. The movement of the imagination is a reflective and self-assertive enterprise. It exhibits, as Fichte notes in the practical part of the *Wissenschaftslehre*, a centripetal and centrifugal oscillation that moves between the self's determining and being determined and that attempts to preserve, within that context of opposition, the unity of the self.

Within the derivations that constitute the *Wissenschaftslehre*, the imagination first appears as an "absolute activity that determines a reciprocity."[9] It becomes in due course a more determinate expression of substance itself,[10] what for Fichte represents the interior/subjective component of experience. Fichte, of course, rejects the traditional notion of substance as a substratum or bearer of accidents, although substance does hold together or give coherence to accidents or properties. Since any accident implies its own opposition,[11] substance becomes that image which provides for the unity and integrity of accidents. That is, imagination forms an image or substrate that makes differing accidents thinkable together and at the same time. As Fichte maintains, it is imagination itself, which "by its most wondrous capacity,"[12] holds fast a fleeting accident long enough to compare it with its opposite that supplants it.[13] Fichte writes:

> It is this forever unacknowledged capacity which connects together persistent opposites into an unity—which intervenes between moments which must mutually destroy one another, and thereby sustains both. It is that which alone makes life and consciousness possible as successive temporal series, and it does all of this ultimately by carrying forward, in and by itself, accidents, which have and could have no common bearer, because they would mutually destroy one another.[14]

The function of the imagination to unite opposites is interconnected with the inevitable and irreconcilable circularity that arises with any attempt to explain the connection between subjectivity and objectivity in a one-sided fashion. The circularity between subjective and objective, I and not-I, can also be seen in the relation between the logical functions of causality and substance. Explanations of experience from the side of causality, the side of objectivity—passivity in the I on the basis of activity in the not-I—fail to clarify, as do all realist views, how the I can be aware of its being affected by the not-I, how such a determination can be *for* the I. By contrast, explanations from the side of substance or subjectivity fail, as do idealist views, to demonstrate how the I's determination is due to something other than the I—in short, how representation is of something external, a thing, a world, or nature.[15] One-sided explanations thereby inevitably imply their opposite in a circle that cannot be unraveled, nor should it be, for the ultimate synthetic unity of causality and substance as relations, the unity between subjective and objective, forms what Wolfgang Janke has called a living and dynamic "*Kreislauf.*"[16] The task is not to transcend the circle, for that is impossible. Rather, in beginning with the I which posits itself as determined by the not-I, one is to bring what is thinkable into an ever more restricted circle "until," Fichte writes, "we finally discover the one way to think what is to be thought."[17]

Ultimately, what must be thought is opposition as such, the *Zusammentreffen*—what Heath translates as "clash"—of I and not-I, the sheer engagement between them, or the mutual "incursion" (*Eingreifen*) of each upon the other. As the *Wissenschaftslehre's* dialectic begins to traverse the final arc of its circle, the task is to think opposites in their simple engagement as such.

Finally we find that opposition cannot be transcended; that is, I and not-I cannot be sublated, for there is no final identity between them. We must face their mutually destructive opposition directly and think their engagement. It is here that creative and productive power of the imagination shows itself. Fichte suggests the following example. Posit light (A) and darkness (B) as opposites with the requirement that they both must stand together contiguously, which, of course, they cannot do since each nullifies or negates the other. One cannot posit a boundary or mediator between the two, for any such boundary-point would be identical neither to light nor to darkness. At the mediate point we would have to say that neither is present and that both are present. There can be no interval between light and darkness, so both make immediate and mutually destructive contact at the mediate point. Their engagement (*Eingreifen*) and their meeting together (*Zusammentreffen*) is their own elimination. We must posit both simultaneously, but we can posit neither one. But the imagination, by extending this mediate point to a moment of time, fashions an image of this medium whereby it receives its reality.[18] The imagination must give this mediate point content and extension if we are to think I and not-I, subject and object, in their mutual relation. Fichte writes: "I can accordingly extend [*ausdehnen*] Z [the mediate point in Fichte's example] by the mere imagination, and it must be so extended, if I am to think the immediate boundary of moments A and B."[19] Only by the imagination's oscillation between opposites, its wavering that encircles and embraces opposites by fashioning an image, an *Ausdehnung* (as thoroughly temporal) between them, can subject and object arise in the first place. It is this extension itself, this boundary or point of contact, that enables us to posit subject and object in a sustainable relation. Wilhelm Weischedel suggests the following helpful example.[20] I observe two objects, say, a house and a tree. Obviously, when looking at one, the other does not disappear from my field of vision. This is because I hold myself in something of a hovering pattern between both. By sustaining and holding both the tree and the house in my vision—that is, in a hovering pattern between them—I can form an image or horizon wherein both can be united. In this way I can observe both objects as adjacent to one another. By constructing the image or horizon, I am able to perceive both tree and house without detriment to either one. Tree and house, which oppose one another and which, without a proper horizon for their relation, eliminate one another from my perceptual field, can be united in one sustained view. Although Weischedel's example concerns the production of spatial contiguity, it can be adapted to the union of opposites that is made possible by time. This hovering and encircling pattern that characterizes the activity of the imagination fashions, ex nihilo as Fichte asserts in one of his Jena lectures,[21] the temporal framework within which the encounter between subject and object that constitutes experience and knowledge can take place.

The imagination must reconcile the irreconcilable, unite what cannot be united, and give representation to what cannot otherwise be represented but what nonetheless demands representation. And in its oscillation between incapacity and requirement, the imagination produces an image that allows opposites to be joined if only for a fragile moment. As it hovers above opposites, the imagination, as Fichte describes it, "touches them, and driven back from them, touches them again, and thereby gives them *in relation to itself* (Fichte's emphasis) a certain content and extension which will be shown in due course as the manifold in space and time."[22] This state, as Fichte delineates it, is a state of intuition and what is intuited is an image formed by the imagination.

Yet what the imagination fashions is not simply an intuition as such but time itself as the medium within which intuition can be represented. Here Fichte is utterly faithful to the spirit of Kant's formulation: time is the necessary condition for any possible experience. And yet, the differences between them are radical. Unlike Kant, Fichte derives time from the play of the imagination itself, and it is furthermore time which makes the subject-object, self-other structure possible.[23] Intuition and time arise together—time is not a *given* manifold of pure intuition—as simultaneous products of the imagination. More importantly, time becomes the background of intelligibility within which a representing subject and a represented object can stand in reciprocal relation. It is the intelligible framework, the stable medium within which the image which is the I and what is imaged in and through the I can stand together in relation. Time is the *Urbild*, the horizon of meaning within which the I's experience of objects unfolds. The image that the imagination fashions is thus an intuition and the horizon whose expansion, through further synthetic acts of the imagination, namely, successive moments of time, becomes the temporal background within which further determinations of the intuited object are possible. Fichte does not always distinguish clearly intuition and the moment of time within which intuition takes place, for both are united and arise together in the image. Although both are spawned by the imagination, it is evident that time makes intuition itself possible. In this sense, time thus has a logical priority.

The temporal mode that the imagination fashions is a momentary "now" (*Jetzt*) within which a discrete sensation can occur. But, as Janke indicates, the imagination is fundamentally a "*zeitigend Einbildungskraft*" that allows us to derive the modes of time from the play of the imagination itself. The synthetic activity that arises within the imagination's wavering extends to a before and an after that completes the modes of time. As it oscillates between opposites, and as one opposite vanishes into its counterpart, the imagination keeps open or extends this fading or vanishing and thereby forms the past. Equally it holds open or sustains briefly the one opposite that supplants its vanishing opponent and produces the future accordingly. By the imagination, so Janke maintains, the interplay between departure and arrival (*Solange/Wielange . . . Bis*) are pulled together into an extended Now.[24] If Janke is correct, the imagination's construction of past and future is a reflective activity, as it follows or reflects upon its formation of the present. Life as such, at least for the theoretical I, is lived and maintains itself in the abiding present. Janke writes: "In this manner the theoretical I holds itself in the abiding present. It lives in the lifeform of an intertwined circle [*Kreislauf*], which swings between the activity of the productive imagination and the categorically fixed interplay."[25] We must be careful, however, not to misconstrue Janke's description. The imagination does not construct the present out of the future and the past as though future and past already possessed an independent grounding. Future and past, themselves grounded in opposition in Janke's account, become possible only by reference to the present.

It is the derivative nature of time that ultimately makes T. P. Hohler's argument that time provides the unity for the I in its theoretical and practical capacities unconvincing.[26] Hohler seeks a more phenomenological basis for temporality in the *Wissenschaftslehre* and argues, accordingly, that the unity between theory and practice is precisely the unity of the ecstacies of time. Since striving always presupposes something outstanding, freedom and striving (in which the I goes beyond itself) are futural and concerned with future possibilities.[27] But the I's striving or self-projection is always initiated from its "Was," its having been or *Gewesen* in Heidegger's account, to which Hohler is deeply indebted. Imagination is the temporalizing by which the I constitutes itself. It extends the present between the two horizons of past and future. Fundamental to Hohler's argument, however, is the foundational priority of the future for Fichte. Hohler is close to the quasi-Heideggerean claim that the future, though not, as Hohler insists, as a reified object, is the meaning of the being of the I. For it is the future, as Heidegger had maintained, which enables the I to come to itself as an I. Hohler's phenomenological analysis must accept the givenness of the future, for his analysis attempts to articulate as precisely as possible how the future shows itself as what makes possible the I's self-constitution. For Hohler as for Heidegger, the present is the confluence of past and future. Yet for Fichte, the imagination produces time and the time it spawns is the present moment. Moreover, the present is not fashioned from the past and the future, it is fashioned simpliciter, ex nihilo, as I mentioned earlier. Time comes to be from the original and creative activity of the imagination. It comes to be as a fleeting moment and as thoroughly contingent, and it is its contingency, as Fichte argues in the *Grundriss*, that enables one to derive the other modes of time from the present itself. What Hohler overlooks is the derived, even the contrived, character of all temporal ecstasies. Past and present are possible only in relation to the present.

One possible defense of Hohler would stress that the future is implied in the contingent nature of the syntheses effected by the productive imagination. The image that imagination constructs marks a partial reconciliation of the opposition between subject and object, or the I's self-positing and its positing of an other, but the actual point of combination (*Zusammensetzung*) is not specified. Recall that the third axiom of the *Wissenschaftslehre* is unconditioned as to content. A specific synthesis is thus one of many possible points of contact. The product of the imagination's determining activity is one determination from many possible determinations; that is, an object is determined in a specific manner (e.g., it is red) to the exclusion of other alternatives. Similarly with respect to the self, any self-determination of the I performed by the imagination is one of many possible modes of self-determination. The I that is thereby open to possibilities is an I that is open to the future. But this collapses possibility (à la Kierkegaard's Climacus) and reads contingency as equivalent to Heideggerean possibility, whereas Fichte had something far more traditional in mind.[28]

II

As adequately construed, the imagination is a relation and arises within the dialectic of the *Wissenschaftslehre* as a more specific determination of the category of relation. Understanding the imagination properly requires that we think correctly the relation between the imagination and the interplay that it determines, between itself and what it opposes and relates. Once again Fichte presents the hopeless circularity between idealist and realist descriptions of the I's relation to the interplay that it determines and, as expected, attempts to synthesize their opposition and thus resolve the impasse between the two explanations by tightening the circle which contains them. The series of syntheses at *Sämmtliche Werke*, I, pp. 209–17, are particularly important for our purpose, for it is here especially that Fichte collapses the distinction between self-presentation and the presentation of objects. The opposition between the subjective and the objective becomes an opposition between the infinity and finitude of the I.

From the idealist perspective the activity of the imagination itself determines the interplay of opposites. Here Fichte clearly has in mind Reinhold's principle of representation. The activity of the I posits and opposes both the subjective and the objective and unites both. Only within the I and, of course, by virtue of the I do they become components and are they conjoined. Thus for every representation within the I the imagination determines one thing as subjective, the other as objective. Yet this explanation, like its other, less refined idealist cousins, is incomplete. While it can account for the presence for what is subjective in the I by the I's own self-positing, it fails to explain the presence of what is truly objective. If we reverse the relation, as the realist would insist, the opposing and conjoining of subject and object through the I's activity becomes possible, not from the mere presence of opposites, but from the *Zusammentreffen* in consciousness. This more refined realism explains the objective element within the I simply on the basis of a check (*Anstoss*) on the I's activity, a check that lies outside of the I's activity but forces the I's own self-limitation. The check on the I's outward-going activity requires a limitation of that activity, the representation of something objective to subjectivity, and the conjoining of both.[29] Again the explanation is one-sided since the check on the I's activity cannot explain the I's limitation as a self-limitation. As with the previous opposition between idealism and realism in the *Wissenschaftslehre*, each explanation is thrown back onto its opposite in a circle of determinability which cannot be escaped.[30] As expected, Fichte synthesizes the two one-sided explanations. The I's own activity, its outward striving, and the check upon it mutually condition one another. The check occurs only as a result of the I's activity of self-positing, that is, its striving is reflected back on itself from which self-limitation (and the whole of representation) naturally follows. And the I's self-determination is conditioned by the check on its activity.[31]

This final synthesis fails to untie the knot of circularity, for it harbors the basic opposition that lies at the heart of the theoretical *Wissenschaftslehre* and thus the opposition which threatens the basic unity of the self, an opposition which only the self as productive imagination can think. The activity of the I in its self-limitation, in setting boundaries on its activity, is at once infinite and finite. Its self-limitation presupposes the infinity of its striving, and its self-positing as infinite, that it posits itself as transcending the check, is a self-determination. The I oscillates between its infinite self-projection and its self-reverting reflection from the check. It moves back and forth from the centrifugal nature of infinite self-externalization (*Sich-Entäusserung*) to the centripetally reflective nature of self-determination. The I's oscillation between its irreconcilable infinity and finitude, the wavering which circles its internal self-rending, its "self-losing"[32] and self-maintaining, is the imagination, a spontaneous activity which provides an extended moment of *Streitfrei* stability by extending the state of the I to a moment of time.[33]

There is an interconnection between self-positing activity and the I's self-determination, between its *self*-determination and its being determined (and thereby finite). The attempt to think the I as such discloses the *Widerstreit* that defines the I's identity. The circle of determinability that characterizes the *Wissenschaftslehre* completes its revolution to the first *Hauptsatz* of the theoretical *Wissenschaftslehre*: the I posits itself as determined. To think the I *as* self-positing, as infinite, is to think the I as determinate. The I that posits itself as self-positing becomes a subject whose own thinking is bounded by an object, even if that object is the self as such. The I's finitude is thus contained already in the I's self-reflection, in the requirement that the I understand its absoluteness or self-positing as constituting its essence.[34] This indicates the initial openness of the I, an openness made possible by the check on its original, pure activity and the resulting reflective activity. To think the I properly is to think the I's self-positing as such and, thereby, to determine oneself as absolute.[35] To think the I as such is already to become a theoretical I; it is to posit oneself as . . . ; it is to determine oneself, to finitize oneself. This is why to think the I is to image the I. Ultimately, self-reflection becomes the product of the productive imagination, for the imagination is precisely how the theoretical I relates itself to what Peter Baumanns calls its groundlessness (*Bodenlosigkeit*).[36]

It is the imagination which attempts to sustain the I, to provide integrity and stability to its inner core, which is otherwise defined by contradiction and dispute. That stability is the product of the imagination's own wavering between infinity and finitude, the wavering that forms the image of the I by extending its state to a moment of time. For the self as a self time becomes the *Urbild* which provides the horizon of meaning for the I's understanding

of itself as an I. It is time which allows the I to see (*sehen*), to borrow a locu-
tion from the 1813 *Wissenschaftslehre*, itself as an I. It is time as such which
is the necessary condition for the coherence of the self.

III

The extension of opposites into a temporal boundary or hermeneutical
substrate which allows both the representation and interpretation of opposites as interconnected is what Fichte calls a state of intuition. As I have
already indicated, Fichte rarely distinguishes between intuition of objects
and intuition of the self as a self. In both cases the imagination makes possible the positing and relating of what is subjective and objective. I intend
to show, however, that the priority of the hermeneutical imagination and
its corollary the hermeneutical function of time becomes evident when
we think through what Fichte means the imagination a "beneficent deception" (*wohltätiges Täuschung*).[37]

What does it mean to say that the imagination is a beneficent deception which makes possible, as Fichte indicates, all of the inquiries of the
theoretical *Wissenschaftslehre*? Does this mean that the images that the imagination projects—sensory objects, time, the self itself—have an illusory
quality?[38] The answer is yes and no, and its ambiguity lies in the ambiguity
of the word image (*Bild*).

At the end of the theoretical part of the *Wissenschaftslehre*, Fichte
reminds us that "all reality, as it is understood for us and as it is to be understood in no other way than in a system of transcendental philosophy, is
brought forth merely by the imagination."[39] By a received (*bekommt*)
reality, therefore, opposites become intelligible as opposites.[40] The reality
brought forth by and received from the imagination is an image. An image,
though a substrate in which opposites can be thought, is not a static copy.
It is not a *Nachbild* which merely mirrors what otherwise exists independent of it. Rather, an image is a dynamic, functional unity. The imagination itself is a dynamic, circling movement which itself has no fixed standpoint. The image it fashions is equally tentative and unstable, similar perhaps in this respect to a Whiteheadean occasion. That is, an image is the
construction of an entire field of reference or field of vision which makes
possible experience itself. The pulling together of an image is a self-con-
struction, in the Whiteheadean sense, by the imagination.[41] Image is only
a function, a horizon through which reality can be seen. The ambiguity in

the concept of image lies in its dual meaning as functional, dynamic, imaging, and as product or image.

The deceptive nature of the imagination coincides with the contrived nature of the physical world in the *Wissenschaftslehre*. For Fichte, the natural world itself is not fundamentally real, but serves simply as the inescapable theatre for moral self-development. And Fichte makes no attempt to soften that view in the final pages of the theoretical *Wissenschaftslehre*. Ultimately, as Fichte himself suggests,[42] only the imagination is a genuine *Faktum* of consciousness. Other facts of consciousness are simply artificial facts of philosophical reflection, but this one fact, what Fichte calls the "original fact," actually corresponds to something present within human spirit and present independent of reflection. What then is present apart from reflection but the imagination, the movement of thought itself?

The imagination is a dynamic function, a wavering and circling movement which fashions the objective world and time as the intelligible framework for that world, in order to give itself a temporary anchor for the struggle for self-coherence and self-understanding. The imagination is world-forming, world-imaging, yet, as Weischedel suggests, always in the service of the I's self-assertion, its attempt "to become its own lord."[43] The external world presented within consciousness, and time as the condition for its possibility, provide an ever expanding and contracting horizon that gives rootage to the finite self and provides the framework within which to work out the basic inconsistency of its nature. It is in this sense that the hermeneutical function of the imagination is more basic than its scientific/epistemological counterpart. In the theoretical *Wissenschaftslehre* Fichte never distinguishes clearly self-presentation and presentation of objects, but it is evident that he primary concern is the self's understanding of itself and its effort to reconcile its dual nature. The dialectical impasse of the theoretical *Wissenschaftslehre* denotes an aporia within the I created by the ineradicable opposition between its infinity and its finitude, a aporia that Fichte attempts to fill with ethical action. But it is time, projected by the imagination, which secures the meaningful framework within which practical striving can take place. Ultimately, it is the imagination alone which is not itself a deception. Fichte writes: ". . . it, however, does not deceive, but offers truth and is the one possible truth. To assume that it deceives is the foundation of a skepticism which instructs one to doubt one's own being."[44] The imagination does not deceive precisely because it is the dialectical movement of thought itself. The thinking of the I is the imagination, a thinking which makes its doubting practically self-refuting.

Yet finally, it is the I itself that is real, and real as the imagination by which the I attempts to give coherence to its nature, a coherence which is aided by the freedom of the self's true other, that is, individual I's in a community of free spirits.

NOTES

1. *Johann Gottlieb Fichtes Sämmtliche Werke*, ed. I. H. Fichte (Berlin:Viet & Co., 1945-46), I, p. 208. Reprinted, along with *Johann Gottlieb Fichtes nachgelassene Werke* (Bonn: Adolphus-Marcus, 1834-35), as *Fichtes Werke* (Berlin: de Gruyter, 1971).

2. "Spirit as such is what is otherwise called productive imagination." Daniel Breazeale, ed. and trans., *Fichte: Early Philosophical Writings* (Ithaca, N.Y.: Cornell University Press, 1988), p. 193.

3. Take, for instance, the category of efficacy (*Wirksamkeit*) or causality. In intuiting an object the I unites things that are opposed while losing itself in its initial intuiting action. In so doing, the I transfers to the object of its intuition something, namely its own activity, which lies in the I. With this transference the category of causality arises, and the synthesis that makes the transference possible is one performed by the productive imagination. Thus, causality as a category of thought originates within the imagination. Categories thus arise together with objects of experience as products of the creative capacity of the imagination. See J. G. Fichte, *Grundriss des Eigenthümlichen der Wissenschaftslehre*, in *Sämmtliche Werke*, I, pp. 386–87. "In the *Wissenschaftslehre* they arise together with objects, and, in order to make objects initially possible, they arise from the soil of the imagination itself." *Sämmtliche Werke*, I, p. 387.

4. The fact that the imagination is both presupposed and derived corresponds to the unavoidably circular nature of the *Wissenschaftslehre* as announced in part 1. Any reflective procedure must presuppose the laws of logic or reflection even in order to derive those same laws. The same applies to the imagination. As a spontaneous activity that enacts synthetic unities, it is presupposed initially and then subsequently derived.

5. Rudolf Makkreel, "Fichte's Dialectical Imagination," in *Fichte: Historic Contexts/Contemporary Controversies*, eds. Daniel Breazeale and Tom Rockmore (Atlantic Highlands, N.J.: Humanities Press, 1994), p. 11.

6. John Sallis, *Spacings—of Reason and Imagination in the Texts of Kant Fichte and Hegel* (Chicago: University of Chicago Press, 1987), p. 64.

7. Fichte writes in the *Grundriss*:". . . on the occasion of an until now completely inexplicable and inconceivable check on the original activity of the I, the imagination, oscillating between the original direction of this activity and the activity originating from reflection, produces something composed of both directions." Fichte, *Sämmtliche Werke*, I, 331.

8. The *Wissenschaftslehre* attempts to retrace the movement of thought implied within a complete thinking of the I's combining its self-positing and its positing of an other. This requires the I's own self-determination, but in a way that preserves the unity of the I demanded by its original self-positing. For Fichte, categories are modes of synthesizing activity (*Handlungsart*) performed by the imagination. They include theoretical and practical acts. As such, they are more determinate types of positing, that is, positing as combining. As Baumann's writes, the categories "originate . . . from the [I's] dialectical, original structure of a theoretical-practical-pre-disjunctive self-positing and counter-positing of a not-I with the principle of striving toward an opposition-free self-positing as a principle of mediation." Peter Baumanns, "Transzendentale Deduktion der Kategorien bei Kant und Fichte," in *Erneuerung der Transzendentalphilosophie*, eds. Klaus Hammaker and Albert Mues (Stuttgart: Frommann-Holzboog, 1979), p. 67.

9. Fichte, *Sämmtliche Werke*, I, 160.

10. That is, the imagination, *as* an independent activity determining a reciprocity, becomes a more determinate expression of substance. Actually, substance is a determinate mode of synthesis and thus a more determinate act of the imagination.

11. Fichte, *Sämmtliche Werke*, I, pp. 204–205, 224–25.

12. Ibid., pp. 204–205.

13. As we will see, this will become the basis for the derivation of temporality.

14. Fichte, *Sämmtliche Werke*, I, p. 205.

15. The problem of this perpetual circle becomes clear as early as synthesis E of part 2. Prior to E the following circle is disclosed. The not-I limits the I because the I limits itself interiorly, yet the interior I limits itself because the exterior not-I limits it from outside. From the side of causation (the not-I) we cannot explain how an external world influences a knowing subject, while from the side of substance (the positing I) we cannot explain how to connect the I's limiting power to the causal agency of the not-I.

16. Wolfgang Janke, *Fichte: Sein und Reflexion: Grundlagen der kritischen Vernunft* (De Gruyter, 1970), p. 160.

17. Fichte, *Sämmtliche Werke*, I, p. 219. The imagination does not simply encircle and embrace opposites in a single act of composition (*Zusammensetzen*). Its procedure from one synthetic act to another is equally circular, since a single synthetic act cannot eliminate all opposition.

18. Ibid., p. 208.

19. Ibid.

20. Wilhelm Weischedel, *Der Frühe Fichte* (Stuttgart: Frommann-Holzboog, 1973), pp. 65–66.

21. "Concerning the Difference between the Spirit and the Letter in Philosophy," in Breazeale, *Fichte*, p. 193.

22. Fichte, *Sämmtliche Werke*, I, p. 225.

23. ". . . we will, on the contrary, prove the ideality of time and space from the demonstrated ideality of objects. He [Kant] requires ideal objects in order to fill time. We require time and space in order to place ideal objects." Fichte, *Sämmtliche Werke*, I, p. 186.

24. Janke, *Fichte*, p. 158.

25. Ibid., p. 161. This would suggest that the categories or syntheses they represent are themselves determinations of time. Fichte never works out within the *Wissenschaftslehre* a typology for the temporal schematization of the categories. This is perhaps because of the more radically creative function of the imagination.

26. T. P. Hohler, *Imagination and Reflection: Intersubjectivity: Fichte's Grundlage of 1794* (Martinus Nijhoff, 1982).

27. Note that Hohler's view takes *Streben* to refer to a nonderivative future. Hohler seems to think that since striving is directed to what is still indeterminate and thus to possibility, and, following Heidegger, that futurity underwrites possibility, the I's most basic activity is related to the future. But this surely is to read Heidegger's analysis of temporality in *Being and Time* back into the *Wissenschafts-lehre*, and to assume incorrectly that Fichte subscribed to a nontraditional view of time, a view clearly inconsistent with the derivation of the past in the *Grundriss*.

It is also not completely clear in Hohler's scheme whether future possibilities are located within the I or not-I. Heidegger's view (and Fichte's as well) would demand that possibilities are implicit within the I's structure. After all, for Heidegger self-understanding is precisely the projection of Dasein's possibilities, and interpretation is simply the laying out (*Aus-legung*) of those possibilities so that they may be clearly seen and appropriated.

28. Fichte's derivation of time in the *Grundriss* is the derivation of a temporal series, that is, time conceived traditionally as a series of synthetic points with each synthetic point as a contingent synthesis. As contingent, each synthesis is dependent upon another point that is thus necessary in relation to it. But that point is also contingent and dependent upon a further point necessary in relation to it, ad infinitum. In this sense we have the emergence of a temporal series. Note, however, that the series is constructed out of present and past moments. The contingency of the present is a contingency in relation to a past moment. The present is such that any other perception, for instance, could have occurred within it. Yet this can occur only if there is another moment wherein one and only one possible perception could have occurred, and this, so Fichte maintains, is the characteristic feature of the past.

Consciousness is thus the awareness both of freedom and identity, identity because there is never consciousness of a discrete moment. Perception B, for instance, cannot be a perception unless one presupposes that the experiencing subject has had another perception A and so on. Fichte stresses that this rule lies at the basis of the identity of consciousness, and—not particularly—only two moments are required for the identity. The present moment of consciousness is then actually always a second moment. "There is no first moment of consciousness, but only a second." See Fichte, *Sämmtliche Werke*, I, pp. 408–10.

29. The check registers in the reflecting I as feeling. It is the task of the imagination to bring feelings to consciousness.

30. Fichte, *Sämmtliche Werke*, I, pp. 209–10. The inescapable circle of determinability appears as an antinomy which can only be overcome by a decree of reason which the imagination enacts and executes.

31. Ibid., pp. 212–14.

32. This is Baumanns's interpretation of *Sich-Entäusserung*. See Baumanns, *J. G. Fichte: Kritische Gesamtdarstellung seiner Philosophie* (München: Verlag Karl Alber, 1990), p. 88.

33. Fichte, *Sämmtliche Werke*, I, pp. 215–17. Michael Vater has suggested to me that Fichte's syntheses and derivations in the theoretical *Wissenschaftslehre* are deconstructive, that is, designed in part to create the very aporia that can be filled only by practical activity, but an aporia held tenuously by the imagination. Baumann's also writes that "productive imagination is thus the despairing (*verzweifelte*) toil of the I to unite what can't be united." (Baumanns, *J.G. Fichte*, p. 89). The stability that the imagination provides in the production of temporality is a stability that temporarily closes the aporia within the nature of the self as such. The entire derivation of the structure of experience is the effort of the imagination to resolve or mend, if for a fleeting moment, the self-rending that constitutes the opposition between the I's self-positing and reflective activity.

34. See Weischedel, *Der Frühe Fichte*, p. 62.

35. This corresponds to the I as image and its self-recognition as an image of an image in the 1813 *Wissenschaftslehre*.

36. Baumanns, *J. G. Fichte*, p. 84.

37. For Fichte's discussion of the deceptive quality of the imagination in relation to the thought of Maimon, see the *Grundriss*, in *Sämmtliche Werke*, I, pp. 387–89.

38. Note the clear connection to Hume, though it is likely that Fichte's sole access to Hume was filtered through Kant, Jacobi, Maimon, and others.

39. Fichte, *Sämmtliche Werke*, I, p. 227.

40. Ibid., pp. 226–27.

41. My brief analogy to Whitehead has benefited from a discussion with Michael Vater.

42. Fichte, *Sämmtliche Werke*, I, p. 219.

43. Weischedel, *Der Frühe Fichte*, p. 68.

44. Fichte, *Sämmtliche Werke*, I, p. 227.

POSITING AND DETERMINING IN FICHTE'S *FOUNDATION OF THE ENTIRE WISSENSCHAFTSLEHRE*[1]

⑦

Günter Zöller

Fichte's project of a transcendental theory of knowledge (*Wissen*), which he developed under the title *"Wissenschaftslehre"* (Doctrine of Knowledge) over a period of some twenty years, eludes classification within the traditional system of philosophical disciplines. The knowledge that Fichte seeks to elucidate in its principal structure includes theoretical as well as practical knowledge, thus pointing to the basis of knowledge of all kinds in a common ground that precedes and prepares its subsequent differentiations. Moreover, Fichte's account of the ground of all knowledge is not only concerned with the forms and structures of knowing. It is as much an account of the principal structure of that which is known in all knowledge. Accordingly, Fichte's transcendental theory of knowledge of all kinds is at the same a transcendental theory of objects of all kinds.

With its comprehensive scope and radical intentions, Fichte's *Wissenschaftslehre* can lay claim to the old Aristotelian title of a "first philosophy." Yet while traditional first philosophy was a theory of the principal forms and kinds of being, Fichte's *Wissenschaftslehre*—following the precedent of Descartes and Kant—takes the form of a theory of the principal forms and the ground of the knowledge of all being. The object of first philosophy in Fichte is not being as such but the knowledge of being, more precisely, the necessary conditions of such knowledge. The epistemic turn in first philosophy, which is characteristic of transcendental philosophy in general, is more than a methodological device designed to start with what is best known. It is based on the substantial rather than merely methodological insight that it is not possible to abstract from one's knowledge or, more generally, one's mental involvement in the consideration of whatever there is and whatever being is. It is this uneliminable presence of principal mental accomplishments which Fichte addresses under the term "I" (*Ich*) in the various elaborations of his *Wissenschaftslehre*.

Given the role assigned to the I as the principal ground of knowledge and its objects, the I in question can not be the empirical self of the concrete human individual. Rather, Fichte's I is a structural complex or a complex structure that precedes and renders possible all individual mental life. Yet much of the terminology and concepts employed by Fichte to address the nature and functions of the I as transcendental principle are borrowed from the sphere of the empirical, concrete I—including the former's appellation as "I" in the first place. This strategy is certainly motivated and probably justified by the consideration that the empirical I is the closest nontranscendental counterpart of the I of transcendental theory or the transcendental I. Still the terminological proximity between the transcendental ground structure and that which it grounds is potentially misleading, did in fact mislead many of Fichte's readers, and was abandoned by Fichte himself in the later versions of *Wissenschaftslehre*.

In Fichte's first and only published detailed presentation of the *Wissenschaftslehre*, the *Foundation of the Entire Wissenschaftslehre of 1794–95*,[2] though, the egological terms and concepts thoroughly pervade Fichte's transcendental theory of knowledge. A distinguishing feature of this presentation is its division into the three Principles of the Entire *Wissenschaftslehre*, on the one hand, and the further development of the complex relations between those principles in the Foundation of Theoretical Knowledge and the Foundation of the Knowledge of the Practical, on the other hand. While the three principles that open the *Foundation of the Entire Wissenschaftslehre* present the most abstract, general traits of all knowledge, the separate but linked treatments of theoretical knowledge and knowledge of the practical, which are based on the interrelations of the three principles, address with growing specificity the principal features of concrete mental life, thus rendering increased plausibility to Fichte's choice of the language of the "I" for his transcendental account of knowledge.

But the relation between the three principles of the entire *Wissenschaftslehre* and the latter's theoretical and practical parts not only documents the crucial transition from the generic conception the I to its principal specifications as theoretical I and practical I. It also documents, in the inverse direction, the origin of the specific forms of knowledge and the forms of mental life associated with each of them in some generic, transcendental ground whose three principal components are responsible for the basic structures of all knowledge and mental life. To use a musical analogy, the three principles of the entire *Wissenschaftslehre* provide the exposition of the themes which are subsequently subjected to a development that brings out their initially latent potential for multiple relations. The *Foundation of the Entire Wissenschaftslehre* has a truncated sonata form.[3]

The complex relationship of prefiguration and unfolding that holds between the ground of the entire *Wissenschaftslehre*, on the one hand, and the grounds of theoretical and practical knowledge, on the other hand, has found linguistic expression in Fichte's judicious employment of the terminologies of *positing* (*setzen*) and determining (*bestimmen*), which pervade the *Foundation of the Entire Wissenschaftslehre*. Both terms and their derivatives are sufficiently generic to address the overall ("entire") structure of knowledge or the I. Yet they are equally susceptible to further specification in order to capture what is peculiar to some but not all aspects of knowledge or the I. Moreover, between them those two terms repeat the developmental tension of the *Foundation of the Entire Wissenschaftslehre*, with the terminology of positing especially dominant in the formulation of the three principles of the entire *Wissenschaftslehre* and the terminology of determining to be found chiefly in the specifically theoretical and practical parts of the work. The four remaining sections of this paper will address Fichte's accounts of absolute positing (I), the transition from positing to determining (II), theoretical determining (III), and practical determining and predicative positing (IV) in the perspective of the unitary structure of the *Foundation of the Entire Wissenschaftslehre* and its object of study, the I.

I. ABSOLUTE POSITING

The principal feature attributed to the I in the three principles of entire *Wissenschaftslehre* is that of *positing*, more precisely, of *positing absolutely* (*schlechthin*). According to the first principle "[t]he I posits originally absolutely its own being" (*Das Ich setzt urspünglich schlechthin sein eigenes Seyn*; 261). The second principle states that the I posits absolutely something over and against itself, namely, the not-I (*Das Entgegengesetztseyn ist überhaupt schlechthin durch das Ich gesetzt*; 266). Finally, as the third principle claims, the I posits absolutely the I as well as the not-I as divisible (*[Es] wird demnach schlechthin das Ich sowohl als das Nicht-Ich theilbar gesetzt*; 270). While all three kinds of the I's principal positing are characterized as absolute, the second and third principal kinds of positing are partially conditioned. The form of the I's op-positing or conterpositing of the not-I is unconditioned (not derivable from the original absolute positing) but the matter or content of the I's counterpositing is conditioned in that the opposed content is the opposite of the posited I, namely, a not-I (266). By contrast, it is the matter or content of the third kind of absolute positing that is unconditioned (the divisibility of I and not-I is not derivable from their being posited and counterposited, respectively), while the form of this kind of positing is contained in the task of reconciling the opposed kinds of the I's positing in the first and second principle (268).

Fichte's choice of the term "positing" to designate the absolute, grounding dimension of knowledge or the I receives no explicit justification in the *Foundation of the Entire Wissenschaftslehre*. Nor is there an explanation of the technical meaning of this term offered anywhere in the work. Since there is also no direct precedent for the specific use of the term in transcendental philosophy before Fichte, it is indicated to interpret Fichte's use of "positing" functionally, as a coinage designed to name a conceptual space opened up in Fichte's transcendental theory of knowledge.

The emergence of the term "positing" in Fichte can be traced to his involvement in the contemporary debate about a philosophy based on a first principle (*Grundsatz*). The conceptual space filled by the notion of positing was first opened up through Fichte's encounter with G. E. Schulze's skeptical critique of K. L. Reinhold's attempt to provide a unitary foundation for Kant's critical philosophy. Under the title "proposition of consciousness" (*Satz des Bewußtseins*), Reinhold had advanced the following formulation of the allegedly self-evident highest principle of all knowledge:

> In consciousness the representation is distinguished by the subject from the subject and the object, and related to both.[4]

In his anonymously published treatise entitled *Aenesidemus* (1792),[5] Schulze had critiqued Reinhold's principle by showing how the latter is not unconditional but both formally and materially conditioned by concepts and principles that it tacitly presupposes.

In his review of *Aenesidemus* (1794),[6] Fichte had responded to Schulze's criticism of Reinhold's version of the first principle of all knowledge by conceding the inadequacy of Reinhold's particular choice of a first principle, while maintaining the overall goal of searching for the first principle and indicating that a true first principle would have to provide the foundations not only for theoretical philosophy but for the "entire" philosophy and would therefore have to employ a more generic concept than the specifically theoretical notion of representation to capture the foundational stratum of knowledge.[7]

In the *Foundation of the Entire Wissenschaftslehre* the fragmentary hints at the pretheoretical, prerepresentational foundation of all knowledge to be found in the *Aenesidemus* review have been developed into a complex account of the I's threefold absolute positing, designed to provide a more promising successor to Reinhold's principle with its distinction and relation between subject, object and representation.[8] The key features of the principal ground of all knowledge (absolute I, absolute not-I, divisible I, and divisible not-I) are presented in separate but related principles, and the term "positing" addresses the generic, transcendental nature the I's principal ground.

Yet while the term "positing," as used to address the absolute structure of the I, has neither a specifically theoretical nor a specifically practical meaning, it nevertheless prefigures the subsequent differentiation of the I into a theoretical I or intelligence and a practical or striving I. Absolute positing in Fichte contains the protocognitive connotation of self-ascription or "taking" as well as the protopractical connotation of activity or doing. Indeed it is the very unity of those two traits that characterizes the I as such. For Fichte the I is originally theoretical and practical. Neither of the two traits is added on to the other. Each can only be what it is in unison with the other.

Still there is a tendency in Fichte to express the complex nature of the I predominantly in terms of activity or doing. But this conceptual strategy is not designed to play down the cognitive or theoretical aspect of Fichte's I, or its prototheoretical antecedent in absolute positing. Rather the main addressee of this move is an unduly reified, static understanding of the I as a thing or object. Fichte seeks to counter such an ontological misunderstanding with an overall practification of the I that includes rather than excludes its cognitive or theoretical dimension. Theory itself is a mode of practice generically understood.

In addition to its prototheoretical-*cum*-practical character, the transcendental ground of all knowledge possesses the essential feature of absoluteness in the I's original positing. The I in its capacity as universal ground is not itself grounded in anything outside of it but is strictly self-grounded. The I's absolute positing is not based on some prior being of the I. Rather it is only in and through the I's absolute positing that the I first comes into existence (259f). In the original I, then, positing and being coincide: the I posits its own being. Yet equally originally the I posits the being of a not-I. The not-I, too, in no way precedes its being posited by the I; it first comes about in and through the I's formally absolute counterpositing.

But just as little as the there is a being preceding the I's positing activity is there a being succeeding this positing as its result or product. The absolute I qua self-posited is nothing apart from this very act of positing. Its being is its being posited and nothing beyond that. Fichte expresses this unique feature of the I that distinguishes it from any other activity through the neologism, "*Thathandlung*" (255, 257, 259, 260)—a philosophical term he coins to go beyond Reinhold's inadequate reliance on facts (*Tatsachen*) of consciousness.[9] Fichte's coinage indicates the identity of the act (*Handlung*) and its product (*Tat*) in the case of absolute self-positing (259). Not even the I's other act of absolute positing, namely, that of the not-I, can count as a case of *Tathandlung*. For in that case the product, the not-I, is not identical with the I's act but different from it and even opposed to the I.[10]

The I that performs the three kinds of absolute positing (self-positing, counterpositing of the not-I, and positing of the I and the not-I as divisible) is the I qua *absolute subject* (260). It is important to distinguish the I in its absolute capacity of preceding all subsequent distinctions with respect to the I from two other forms under which the I appears in the system of supreme principles, namely, the I qua substance that is being divided (by the absolute I's positing) into divisible I and divisible not-I, and the I qua accident that is opposed in the substance I to the divisible not-I (279). Nor is the I qua subject of absolute positing to be identified with either the theoretical I or the practical I which figure in the specific parts of the *Foundation of the Entire Wissenschaftslehre*.

Strictly speaking, the absolute I is not an I at all. On the basis of the three forms of positing alone, there is no consciousness or self-consciousness. The absolute I, like its correlates (substance I, accident I, accident not-I) are merely so many moments that enter into the constitution of actual consciousness and self-consciousness.[11] Thus the formally absolute positing of the not-I does not amount to the consciousness of an object, and the materially absolute positing of divisible I and not-I does not yet establish any of the specific relations between I and world, such as the relation of knowing or the relation of acting. The absolute ground of knowledge is an I only proleptically speaking, in anticipation of the I made possible by those very principles. Yet given the task to articulate the radical heterogeneity of the absolute ground of all knowledge from every thing grounded therein, the terminology and conceptuality of the I with its con-notations of spontaneous, self-regarding activity might still be considered one of the less objectionable choices.

II. FROM POSITING TO DETERMINING

The three kinds of the I's absolute positing provide only the most abstract scaffolding for Fichte's reconstruction of the conditions and forms of all knowledge. In particular, the relation of the three absolute acts to each other receives only the most general characterization. Neither the abstract combinatorics of materially and formally unconditioned, formally uncon-ditioned, and materially unconditioned positing nor the differentiation of the I according to the titles of subject, substance and accident addresses the *specific* features of the I's overall constitution.

The lack of specificity in the initial presentation of the I's activity of positing is due to the absolute nature of that activity. In each of the three original acts the I's positing occurs (totally or in part) absolutely, i.e., without conditions and entirely spontaneously. Even the counterpositing of the not-I is a formally unconditioned act of the I qua absolute subject; similarly, the positing of the I and not-I as divisible is a materially uncon-ditioned act of the absolute I. Their very unconditioned character renders the three absolute acts of the I and the absolute I itself infinite; they do not have boundaries that would make them one thing rather than another.

Yet while the threefold absolute positing of the I is unconditional and infinite, the absolute activity of the I includes the very positing of speci-ficity as such. In the second principle the absolute I posits a not-I, and in the third principle the absolute I posits a divisible I opposed to a divisible not-I. The resultant constellation of substance I, accident I and accident not-I is characterized by the feature of *determination* (*Bestimmung*) or *limita-tion* (*Begrenzung*) (282). The I qua substance provides a totality to be divided up in yet to be specified ways among the accidental I and not-I, which thus stand in the relation of limiting or determining each other (385).

By contrast, the absolute I of the first principle is both undetermined and undeterminable. Strictly speaking, the absolute I is not anything at all (271). It is merely its own positing, which in turn consists in nothing but that very positing of itself. Logically speaking, the absolute subject is without any predicate. Fichte introduces a type of judgment peculiar to the absolute I and its manifestations, namely, the thetic judgment (*thetisches*

Urtheil), in which the position of the predicate determining the absolute subject is left open ad infinitum (116-18).[12]

While the thetic judgment concerning the absolute subject lacks any determinate, finite predicate that would in turn determine the subject, all other forms of judgment stand under the logical principle of ground or reason (*Satz des Grundes*), which is the formal-logical version of the transcendental principle of mutual determination (272). All nonthetic judgments have their ground in a feature (*Merkmal*) that determines the relation between subject and predicate. Fichte distinguishes between two types of nonthetic judgment, synthetic and antithetic (or analytic) judgments. In synthetic judgments the predication is based on a ground of relating (*Beziehungsgrund*) with regard to which what is otherwise opposed to each other is identical. Antithetic judgments involve a ground of distinguishing (*Unterscheidungsgrund*) with regard to which what is otherwise identical is being opposed (272, 278f).

Thus the introduction of determination in the third principle marks the transition from an absolute, infinite I to an I that is finite or limited with respect to the not-I. But the third principle of the entire *Wissenschaftslehre* marks equally the introduction of a limiting or determining I: the I qua accident determines the not-I qua accident just as much as the not-I determines the I. In fact, the precise nature of the mutual relation of limitation or determination between finite I and not-I is the very subject matter of the remaining parts of the *Foundation of the Entire Wissenschaftslehre*.

But that relation can only be established with reference to the I's original infinite, unconditioned nature. Divisible I and not-I are not equal partners with an equal claim to limiting or determining the other. Rather, the I's original absolute status provides the directional sense for the I-not-I relation: Just as the absolute I is the source of all positing, so the finite I is determined to be the one that is determining rather than the one that is being determined in the I's relation to the not-I (403f). The term "determination" here takes on a finalist connotation, thus creating a telling finitist-finalist double meaning, which permeates much of Fichte's thinking on the determination or destination of the human being (*Bestimmung des Menschen*).

III. THEORETICAL DETERMINING

The mutual limitation or determination of I and not-I introduced in the third absolute principle divides into two half-statements, each of which functions as the principle for one of the two specific parts of the *Foundation of the Entire Wissenschaftslehre*. Either the I posits itself as determined by the not-I, or the I posits itself as determining or limiting the not-I (*das Ich setzt sich, als bestimmt durch das Nicht-Ich; das Ich setzt sich als bestimmend das Nicht-Ich*; 385). The first is the case of the theoretical relation, in which the not-I is supposed to bring about determinations in the I; the second is the case of the practical relation, in which the I is supposed to bring about determinations in the not-I.[13] Fichte postpones the analysis of the practical relation between I and not-I until after the treatment of the theoretical relation, arguing that any determination of the not-I by the I presupposes the reality of the not-I, which can not be taken for granted at this point and would first need to be established in the theoretical part of the *Wissenschaftslehre* with its derivation of the not-I as source of the I's being determined (285).

As it turns out, the analysis of the principle of theoretical determination does not result in establishing the absolute, independent reality of the not-I. The not-I remains something posited by the I. But there is a more limited reality to be accorded to the not-I, which can become the focus of the analysis of practical determination of the not-I by the I.

Moreover, not even the first half-statement of the third absolute principle will stand up to closer scrutiny. The notion that the I is determined by the not-I stands in flagrant contradiction to the basic understanding of the I as absolutely positing. After all, it is the I itself, in its capacity as absolute subject, which posits the not-I's determining of the I. And while the I that is determined by the not-I is not the I in its absolutely positing capacity, the I subjected to determination by the not-I is still an I with its characteristic feature of spontaneous activity. Otherwise the Not-I would not be determining an I but another not-I.

The task of the theoretical part of the *Wissenschaftslehre* is therefore to mediate between the apparent passive status of the I as the recipient of determinations through the not-I, and the absolute nature of the I. Fichte introduces an elaborate apparatus of mediations between the absolute and the theoretical functions of the I, designed to progressively eliminate the contradiction between absolute independence and passive determination (283–85). While this process manages to minimize the contradiction underlying theoretical determination, the contradictory opposition between the absolutely active and the merely passive moment in the I never vanishes completely.

Fichte correlates the successive attempts at eliminating the fundamental contradiction inherent in theory as such to a series of philosophical standpoints (309ff). Those range from extreme, "dogmatic" realism, defending the total determination of the I through the absolute reality of things in themselves, to more moderate forms of realism and idealism that seek to balance the active-passive double nature of the I with the attribution of various degrees of independence to the not-I. The point of this systematic critique of philosophical positions is to generate by way of negation the one standpoint on theoretical knowledge that does justice to the I's spontaneity and independence while preserving the insight that in the theoretical relation the I undergoes determination (279f).

In rough outline, Fichte's solution to the basic contradiction of the theoretical relation is twofold. For one, he reduces the role of the not-I from that of the source of all determinations in the I to that of a mere "check" (*Anstoß*) that sets in motion a process of self-limitation or determination by itself on the part of the theoretical I (355f). The not-I is not actually determining the I nor is the I being determined in specific ways. Rather the I is determined in a most general, unspecific way to bring about its own determinations at the instigation of the "check." Any specific determination of the I is entirely the work of the I's own activity. In the end, the not-I is a product of the positing I (361), and the check merely provides the occasion for the I's self-limiting activity.

Moreover—and this is the second part of Fichte's solution of the theoretical contradiction—the theoretical activity of the I, while originating entirely in the I, is not always present to the I. Fichte introduces a sharp distinction between the philosophical consciousness of the I's monopoly on determination and the ordinary consciousness of the I as always also determined by something other than the I. Only in sustained philosophical reflection does it become clear that even in the case of the theoretical relation the I is strictly speaking entirely self-determining (except for the minimal influence of the not-I under the form of the check)—albeit unbeknownst to the ordinary I involved in the theoretical relation. What *appears* to the theoretical I as the not-I determining the I is actually, as seen by the philosopher, the I's being both what does the determining and what is being determined in a self-relation of determination or relation of self-determination (371). The apparent interaction of I and not-I turns out to be a circle-like interaction of the I with itself (245f).

IV. PRACTICAL DETERMINING AND PREDICATIVE POSITING

Since the theoretical part of the *Wissenschaftslehre* has not resulted in establishing the absolute reality of a not-I that determines the I, the practical part cannot presuppose an original I-independent reality to be determined by the I, either. Rather the examination of the practical relation between a determining I and a determined not-I must itself derive or, in Fichte's preferred parlance, "deduce" the very reality of the not-I. The account of theoretical knowledge in the second part of the *Foundation of the Entire Wissenschaftslehre* presupposes the the knowledge of the practical in the work's third part (286).

Given the absolute, independent nature of the I in general, the not-I can only exercise determination upon the I qua theoretical I, if the absolute I has previously determined the not-I. As in the case of theoretical determination, the I determines itself. But now, in the case of practical determination, the I's self-determination occurs indirectly, with the not-I mediating between the absolute I which determines it and the intelligent I which it determines. The I's practical determination of the not-I is a "detour" (*Umweg*; 389) within the I's overall itinerary of circulating strictly within itself (387f).

The question arises, though, what brings about this detour through the practical in the first place. More specifically, one wonders how the I qua absolute subject comes to "stay open" (this is Fichte's metaphor [405]) for something other than its infinite self-positing. After all, at this point of the transcendental reconstruction of the I and its knowledge a recourse to a preexisting not-I is not available. Any presence of the not-I in or to the I, such as the "check," already presupposes the I's openness to such a "touch." The same impasse seems reached that halted the limitation of the basic contradiction in the theoretical sphere.

Fichte undertakes several attempts at answering the crucial question how the I comes to limit or determine its own absolute and infinite activity. At one point, he refers to everyone's own experience of something alien (*Fremdartiges*) in us that remains quite unspecific but announces something other than the I in the I itself (400). He also introduces feeling (*Gefühl*) as the factual, undeducable source of the I's determinateness. On the basis of the felt, subjective state of being determined the I is then supposed to posit a world of objects as the conditions of the possibility of such feeling (401, 419). Most importantly, though, Fichte argues that it is an essential feature of the I itself, namely, its nature as intelligence, which brings about finitude and determination in an otherwise absolute activity. In order for there to be an I in the fully functional sense of the term, the activity *of* the I must also be an activity *for* the I. Fichte explains the difference between the two levels of the I's activity by distinguishing between the I's "positing itself for some intelligence other than itself" (*sich . . . selbst setzen für irgendeine Intelligenz ausser ihm*) and the I's "positing itself as posited by itself" (*sich setzen, als durch sich selbst gesetzt*; 409).

Fichte characterizes the positing-as, which is required for the full functioning of the I, as "retrieval" (*Wiederholung*) of the original absolute positing (409). The relationship between first, retrieved positing and second, retrieving positing is not one of mere duplication. The second positing does not repeat the first positing but recuperates it, and this in such a way that the I which is posited absolutely by itself is now posited *as* such an I. While the original absolute positing of the I by the I occurs thetically, the second, retrieving positing has the form of predication. More precisely, the second positing is autopredicative: the I posits itself as self-posited. In all other predication, which is predication of something other than the I itself, the retrieval does not concern the I that posits itself but the not-I that is posited by the I and which in turn is posited as such.

Fichte calls the predicative self-positing of the I its activity of "reflecting on itself" (*über sich selbst zu reflektieren*; 408). "Reflection" is to be understood as the I's constitutive trait of attending to its own activity. The I's reflection on itself in predicative self-positing is not a case of positing yet again, and for the second time, some already preexisting I. Rather the positing of the I as self-posited first brings about an I which knows itself as such. Thus the original reflection on the self-posited I is productive and itself a case of the I's original positing activity.

Now this essential reflective activity of the I provides the very occasion for the occurrence of determination in the I. Through the positing of something *as* something, which is peculiar to predicative positing, there enters into the originally undetermined self-positing activity of the absolute I the feature of being determined *as* some such self-positing activity. And while the absolute status of the I might suggest that the I is found to be not just something but everything, namely, all of reality, that expectation has already been undermined by the very fact of the I's own counterpositing. The I that posits itself in original reflection is not a perfectly infinite being or God, but an imperfectly infinite (and imperfectly finite) being, one that is finite in its infinity and infinite in its finitude.[14]

In a perfectly infinite being, whose status as all-encompassing substance eliminates any distinction between that which is being reflected upon and that which reflects, no self-consciousness of the kind to be found in human consciousness would be possible (275). Only a determined intelligent being, i.e., an intelligence that is something but not something else—which possesses some determinations but lacks others—has the determinateness as which it posits itself in original reflection.

Yet reflection can bring about such consciousness of determinateness only if that which is determined as such by reflection already possesses determination, more precisely determinateness. With respect to the determinateness of the I that is presupposed by original reflection Fichte distinguishes between practical and theoretical determinateness. The practical determinateness of the I consists in the sum of all that is given to the I as a task under the form of the "series of the ideal." By contrast, the theoretical determinateness of the I consists in the givenness of the check, which involves the limitation of the I thorough the "series of the real" (409f). The two sides of the original reflection of the I correspond to the fundamental double character of the I as practical-ideal and theoretical-real.[15] With regard to both series Fichte returns once more to the thought of some irreducible and nondeducible givenness of elementary determination—practically as command or obligation and theoretically as check.

The introduction of predicative self-positing through original reflection as a principal feature of the I's positing activity at the end of the *Foundation of the Entire Wissenschaftslehre* reveals the uneliminable gap between the actual character of the I as finite or determined (more precisely, as self-determined in a "checked" way), on the one hand, and the notion of the I's absoluteness, which is now reduced to the status of an idea to be strived after, on the other hand (408f). In its capacity as practical I determining or limiting the not-I, the I strives to assimilate everything not-I to the I—in an attempt to realize the originary absolute nature of the I under conditions of finitude as revealed in counterpositing, check and feeling. Yet the practical I remains forever striving. There is no returning to the unconditioned status of I expressed in the first principle. In fact, the I seems to never have had that status in the first place. And whatever held that status was certainly not an I.

Where does that leave the practical determination of the I through the not-I? To be sure, there is progress. But it is infinite (410). The positing of the not-I and its manifestations as check and feeling are so much part of the I's own constitution that they could only be worked off at the price of self-destruction. Moreover, the finite changes that are actually brought about by practical determination are ultimately only changes in the I's self-determination resulting in a different way of positing the not-I. Our acting, even if it contributes to practical progress, is "ideal," i.e., it takes place for us through the medium of representation (451).[16] But then again, the determining of the not-I is no more and no less real than the very positing of not-I. All reality, whether theoretical or practical, is posited by the I. Fichte was not to address what is ultimately real—beyond all positing and counterpositing, determining and being determined—until several years after laying the foundation of the entire *Wissenschaftslehre*.

NOTES

1. Work on this essay was supported by a Faculty Scholarship from the University of Iowa and carried out during my association with the Center for Literary and Cultural Studies, Harvard University, in the spring of 1995.

2. *Grundlage der gesammten Wissenschaftslehre, als Handschrift für seine Zuhörer.* The *Foundation of the Entire Wissenschaftslehre* will be cited (in the main body of the paper) and quoted from *J. G. Fichte-Gesamtausgabe*, eds. R. Lauth and H. Gliwitzky (Stuttgart/Bad Cannstatt: Frommann-Holzboog, 1962ff.), series 1, vol. 3. All translations are my own.

3. An example of a work of music in sonata form that lacks the reprise section would be Beethoven's second *Leonore* Overture (op. 72).

4. Karl Leonhard Reinhold, *Beiträge zur Berichtigung bisheriger Mißverständnisse der Philosophie* (Jena: Manke, 1790), p. 167.

5. Gottlob Ernst Schulze, *Aenesidemus order über die Fundamente der von Herrn Professor Reinhold in Jena gelieferten Elementar-Philosophie* (1792; reprint, Berlin: Reuther und Reichard, 1911).

6. "Recension Aenesidemus," in *J. G. Fichte-Gesamtausgabe der Bayerischen Akademie der Wissenschaften*, eds. Reinhard Lauth, Hans Gliwitzky, and Erich Fuchs (Stuttgart–Bad Cannstatt: Frommann-Holzboog, 1964ff), I/3: 41–67.

7. Ibid., I/3: 43.

8. On the relation of the *Foundation of the Entire Wissenschaftslehre*, especially its Principles section, to Reinhold Elementary Philosophy, cf. Robert Lawrence Benson, *Fichte's Original Argument* (Ph.D. thesis, Columbia University, 1974).

9. On the relation between *Tatsache* and *Tathandlung* in Fichte, cf. Jürgen Stolzenberg, "Fichtes Satz 'Ich bin.' Argumentationsanalytische Überlegungen zu Paragraph 1 der *Grundlage der gesamten Wissenschaftslehre* von 1794–95," in *Fichte-Studien* 6 (1994): 1–34. On the relation between *Tathandlung* and absolute I, cf. Lore Hühn, *Fichte und Schelling oder: Über die Grenze des menschlichen Wissens* (Stuttgart-Weimar: Metzler, 1994), pp. 77–100.

10. On the relation between *Tathandlung* and intellectual intuition—a term that, while conspicuously absent from the *Foundation of the Entire Wissenschaftslehre*, figures prominently in several of Fichte's other works from that period—cf. Jürgen Stolzenberg, *Fichtes Begriff der intellektuellen Anschauung. Seine Entwicklung in den Wissenschaftslehren von 1793–94 bis 1801–02* (Stuttgart: Klett-Cotta, 1986).

11. On the status of the absolute I as moment rather than independently functioning entity, cf. Peter Baumanns, *Fichtes ursprüngliches System. Sein Standort zwischen Kant und Hegel* (Stuttgart-Bad Cannstatt: Frommann-Holzboog, 1972), pp. 168f.

12. On the systematic function of the thetic judgment in Fichte, cf. Wolfgang Janke, *Vom Bilde des Absoluten. Grundzüge der Phänomenologie Fichtes* (Berlin-New York: de Gruyter, 1993), pp. 204–12. For a critical assessment of Fichte's thetic judgments, cf. Peter Baumanns, *J. G. Fichte. Kritische Gesamtdarstellung seiner Philosophie* (Freiburg-Munich: Alber, 1990), pp. 78–82.

13. Fichte's repeated use of the modal verb *sollen* in his discussion of the relationship of determination between I and not-I is not an expression a command or obligation but indicates indirect discourse.

14. On the distinction between the absolute I and God, cf. Max Wundt, *Fichte-Forschungen* (Stuttgart: Frommann-Kurtz, 1929), pp. 265–80.

15. Concerning the ideal-real duplicity of the I for Fichte, see the author's "L'idéal et et réel dans la théorie transcendantale du suject chez Fichte: Une duplicité orginaire," in *Cahiers de Philosophie* (1995).

16. On the idealist nature of Fichte's theory of action, cf. my "Changing the Appearances. Fichte's Transcendental Theory of Practical Subjectivity," in *Proceedings of the Eighth International Kant Congress. Memphis 1995*, ed. H. Robinson, vol. 1 (Milwaukee: Marquette University Press, 1995).

COMPARATIVE STUDIES

SATZ AND URTEIL IN KANT'S CRITICAL PHILOSOPHY AND FICHTE'S GRUNDLAGE DER GESAMTEN WISSENSCHAFTSLEHRE

Jere Paul Surber

INTRODUCTION

Most readers of Fichte today are inclined to take his occasional protestations that his system is "none other than the Kantian"[1] with at least a few grains of salt. Certainly the "external features" of Fichte's early *Wissenschaftslehre* bear little structural resemblance to either Kant's Critical writings or to their further "systematic" development in the "Metaphysic of Nature and Morals." While a number of commentators have discussed in some detail the differences between Kant's and Fichte's notions of "system,"[2] one finds fewer discussions of the deeper-seated structural and terminological shifts from which their differing conceptions of system arise. While comparisons of Kant and Fichte at the broader level of systematic reflection provide a helpful orientation, we will not be able adequately to assess the degree of their divergences until we have, as it were, "looked under the hood" at the different ways in which in their actual "machinery" functions.

Of course, one must grant that, as Fichte himself was certainly aware,[3] the general idea of a philosophical system stands in a relation of strict reciprocity with its operative terms and statements, so that one can never be considered as entirely independent of the other. Still, before this interconnection between systems and their concrete articulations can be worked out, it is important to be aware of operative differences at both "levels."

In this essay, I propose to adopt the narrower focus in an attempt to chart a crucial shift from Kant to Fichte in the "deep structure" of their respective philosophical orientations. Specifically, I want to claim that Fichte, in effect, reverses the assumed relation between *Urteil* and *Satz* upon which Kant's project had relied and that this reversal had very significant consequences for the divergence of his own systematic enterprise from that of Kant. As my discussion proceeds, I want to pay particular attention to the way in which Fichte's use of the term *Satz* implies a sensitivity to the importance of broadly linguistic issues in the articulation of a philosophical system, something not only absent but (as we will see) explicitly rejected by Kant himself.

I. KANT'S VIEW OF THE RELATION BETWEEN *URTEIL* AND *SATZ*

One can turn to Kant's lectures on logic for a definition of the former term as well as an important *Anmerkung* concerning what he regards as the proper way to understand its relation to the term *Satz*. I will begin by quoting the definition in question.

> A judgment [*Urteil*] is the representation of the unity of the consciousness of various representations or the representation of their relation insofar as they constitute a concept.[4]

Kant's *Anmerkung* relevant to this runs (in part) as follows:

> It is upon the difference between problematic and assertoric judgments [*Urteile*] that the true difference between *Urteile* and *Sätze* is based, a distinction which ordinarily has to do with the mere expression through words without which one would in general not be able to judge at all. In *Urteile* the relation of the various representations to the unity of consciousness is thought merely as problematic, while in a *Satz*, by contrast, (it is thought) as assertoric. A problematic *Satz* is a *contradictio in adjecto*. Before I can have a *Satz*, I must first judge; and I judge about much with which I might not concur, even though I must do so as soon as I take an *Urteil* to be a *Satz*.

Several points deserve note here.

(1) For Kant, the notion of *Satz* is clearly logically subordinate to that of *Urteil* in that he explicitly identifies *Satz* with a specific form of

Urteil, namely, the "assertoric judgment." That is, a *Satz* affirms as actual or true the "content" of an *Urteil* that can otherwise be entertained as merely possible. In so doing, the speaker of a *Satz* is committed to additional consequences, which do not follow if the judgment is expressed in a merely "problematical mode."

(2) Here Kant acknowledges that, in common usage, *Satz* often has a grammatical or linguistic sense (which might tempt us to translate it as "sentence" or "statement"). However, he wishes to exclude such an interpretation from his own "logical" use of the term, where it indicates merely a species of judgment.

(3) Referring back to Kant's definition of *Urteil*, we can see why Kant would want to introduce such a restriction on the meaning of *Satz*. For Kant, a judgment is a "representation," either of the unity of other representations or of their relation to a concept (which, given the way in which he usually defines "concept," amounts to the same thing). As a "representation" (*Vorstellung*) of the unity of other "representations," a judgment is dependent upon its "representational contents," which must first be available before we can judge either "problematically" or "assertorically" with respect to them, that is, before we can regard them under the modality of either possibility or actuality. Although Kant, here and elsewhere, clearly wants to avoid any "psychologizing" of logic, his treatment of judgment as a form of "representation" and his relegation of *Satz* to a species of judgment renders this virtually impossible.[5]

(4) This discussion immediately refers us to another "definition" which Kant offers here, that of *Grundsatz*.

> Immediately certain judgments *a priori* can be called *Grundsätze* insofar as other judgments can be derived from them, but they themselves can be subordinated to no others. On account of this, they are also called Principles [Beginnings].[6]

Note that Kant clearly suggests that there can, in principle, be a number of *Grundsätze*. Indeed, in the *Critique of Pure Reason*, this is precisely the term he most frequently uses for his "synthetic principles of the pure understanding."[7]

Viewed from a somewhat broader perspective, we can put the matter in this way. Given Kant's discussion of judgment and his stipulated limitation of the word *Satz* to indicate the species "assertoric judgment," we can say that the primary meaning of *Satz* for Kant fluctuates (one would like here to use the wonderful German term *schwebt*) between what would (assuming a now-familiar set of distinctions) be expressed in English as, on the one hand, a "proposition" and, on the other, a type of mental act by which its "propositional content" is affirmed. Note, however, that in no case, on Kant's account, should *Satz* be translated as either "sentence" or "statement," since this would be the sort of rendering which Kant himself wishes explicitly to exclude. On the other hand, so long as we are not misled by the homophony of the English term "principle" and Kant's own use of *Prinzip*, we might defend certain current translation practices, admitting that "principle" is beset with the exactly the same ambiguity as the term *Satz* as Kant defines it.

It should be noted that in the background of all of this stands Kant's firm insistence, clearly stated early in section 39 of the *Prolegomena to Any Future Metaphysic*, that logical and grammatical considerations are completely antithetical to one another. Since he regarded any grammatical or linguistic inquiry as a purely empirical matter, Kant categorically denied that such considerations had any relevance whatsoever for a philosophy concerned with "logical/transcendental" issues and banished them wholesale from his philosophical domain. His treatment of the term *Satz* in his lectures on logic is completely consistent with this attitude. As we will soon see, however, this view was by no means shared by Fichte.

II. FICHTE'S REVERSAL OF THE RELATION BETWEEN *URTEIL* AND *SATZ*

Kant, we may recall, claimed, in the introduction to the second edition of the *Critique of Pure Reason*, that "the proper problem of pure reason is contained in the question: How are synthetic judgments possible *a priori*?"[8] Given the oft-repeated centrality of this question about *Urteile* in Kant, what begs for explanation is both the absence of any discussion whatsoever about *Urteil* in Fichte's *Ueber den Begriff der Wissenschaftslehre* and the fact that he only takes up this issue in the *Grundlage der gesamten Wissenschaftslehre* a full two-thirds of the way through the work.[9]

To understand why Fichte proceeded in this quite "un-Kantian" manner, let us begin by considering Fichte's important summary description of the task of the *Wissenschaftslehre* in "*Ueber den Begriff* . . . ."[10] On the reading I am proposing, the heart of this passage is:

Let me make myself even clearer: the *Wissenschaftslehre* establishes a proposition [*Satz*] which has been thought and [then] expressed in words.[11] Such a proposition [*Satz*] corresponds to an action [*Handlung*, here and subsequently] of the human mind [*in menschlichen Geiste*], an action which, in itself, does not necessarily have to have been *thought of* at all. Nothing has to be presupposed for this action—nothing except that without which it would not be possible qua action. This is not something which is tacitly presupposed; it is, rather, the business of the *Wissenschaftslehre* to establish [*aufzustellen*] this presupposition[12] clearly and definitely, and to establish it *as* that without which the action in question would not be possible.

Several points should be noted.

(1) The description of the task of the *Wissenschaftslehre* given here indicates that it involves the articulation, in the form of a *Satz*, of a "mental act." Indeed, nothing other than a *Satz* is mentioned, here or elsewhere in this essay, as the fundamental form of the expression of "thought."

(2) Fichte is very explicit in what he means by a *Satz* here: It is something capable of both being thought and being articulated linguistically (*ein degachter und in Worte gefasster Satz*). Unlike Kant, that is, Fichte explicitly invokes the linguistic dimension of the meaning of this term. Indeed, as this passage is actually written (as opposed to the English translation) one might infer that the thinking of it and its verbal articulation are part of one and the same act or *Handlung*. It is, in any event, by no means clear that Fichte wants to assert that the thought and linguistic expression are necessarily two temporally distinct acts (though he doesn't seem to exclude this either). My suggestion here would be that Fichte is referring to the basic *Satz* of the *Wissenschaftslehre* (i.e. the *Grundsatz*) as a particular sort of "linguistic performative" or "speech act" (which would fit well what has just been said).

(3) When Fichte goes on to claim that this *Handlung* "does not necessarily have to the *thought of* at all" (*die an sich gar nicht nothwendig gedacht werden musste*), I think that this should be read in the context of other of his remarks elsewhere that the *Wissenschaftslehre* is itself a "free construction," or, more specifically, that the "speech act" or "linguistic performative" which constitutes its *Grundsatz* is an absolutely free and hence unconditioned act which has no ground beyond its own spontaneous expression.[13]

(4) As this passage continues, this is further clarified when Fichte draws a distinction between the *thought* directed toward the original unconditioned action and the unconditioned action itself. For the unconditioned action, i.e., what in the *Grundlage* he will call a *Tathandlung* the expression of which is the *Grundsatz*,[14] is presupposed by any subsequent attempt to reflect upon it, but this "act of systematic reflection," while conditioned once it is undertaken, is not itself necessary. It should be noted that, as this discussion proceeds, Fichte continues to employ the schema of *Handlung* and *Satz* as the operative elements in the systematic elaboration which he is describing.

(5) Finally, we should note, in this connection, one of several parallel descriptions of the *Wissenschaftslehre* which occur in "*Ueber den Begriff. . . .*"

> A science possesses systematic form. All the propositions [*Sätze*] of a science are joined together in a single first principle [*Grundsatz*], in which they unite to form a whole.

That is, the *Wissenschaftslehre*, as the "science of science" as Fichte sometimes calls it, is constituted as a unitary totality of *Sätze*, not, be it noted, *Urteile*, and it is these *Sätze* which are, at the same time, the sole medium through which this totality is articulated.

Referring back to our discussion of Kant, then, we must conclude that an important semantical and conceptual shift has occurred in the meaning and "systematic significance" of the term *Satz* when we enter the realm of Fichte's view. In direct contrast to Kant, whose discussion of *Satz* as a particular form of *Urteil* explicitly rules out understanding it in any linguistic sense, Fichte just as explicitly emphasizes its linguistic character and its close connection with the *Handlung* or "mental act" which it concretizes and expresses. For this reason, while we could never fairly translate *Satz* in Kant as "sentence" or "statement," I believe that this must always be counted as a part, if not the central core, of Fichte's use of this term.

III. FICHTE'S TREATMENT OF *URTEIL* IN THE GRUNDLAGE

In order to complete our assessment of the relation between *Satz* and *Urteil* in Kant and Fichte, it remains to consider what place Fichte assigns to *Urteil* in the *Grundlage* and what significance this has for his systematic project in relation to that of Kant. As already mentioned, Fichte does not take up the notion of *Urteil* until almost the end of part 2, which section he calls the "*Erster Lehrsatz*." [15] It is important to notice that he does so in strict connection with his discussion of *Urteilskraft* (which is unfortunately obscured in the English translation of Lachs and Heath, where they use the term "judgment" for both *Urteil* and *Urteilskraft*.)

According to Fichte,

> *Urteilskraft* is the capacity, free until now, to reflect upon objects already posited in the understanding [*im Verstande*], or to abstract from them, and to posit them in the understanding with further determination proportionate to this reflection or abstraction.

To understand Fichte's viewpoint here, it is necessary to consider the context of this very brief section. Fichte takes up the notion of *Urteilskraft* (and hence the *Urteile* which are its expressions) as a sort of transition between his discussion of *Anschauung* (usually, though somewhat misleadingly, translated as "intuition") and *Denken* (thinking). Actually, on Fichte's view, *Urteilskraft* turns out to be merely a counterpart or perhaps even function of *Verstand* (understanding) and is accorded no independent treatment on its own (as opposed, for example, to his discussion, in close lying passages, of *Einbildungskraft*, that is, imagination). Fichte argues here that *Urteilskraft* and *Verstand* reciprocally determine one another. On the one hand, understanding already provides the "objects" (presumably concepts) upon which judgment reflects or from which it abstracts. On the other, judgment determines the "object in general" which is the necessary referent and presupposition for the operation of the understanding. Taken in this reciprocally determining relationship, both together constitute the basis for that which is "thinkable" as any determinate content. His discussion thus issues in a further process of reflection involving the relation between what can in general be thought (any possible "object of thought") and the act of thinking itself.

The most remarkable thing here, I think, is simply the degree to which Fichte demotes *Urteilskraft* and its *Urteile* to an almost "vanishing moment" of his systematic enterprise. Whereas Fichte has a good deal to say about "intuition" on the one hand and "thinking" on the other, "judgment" becomes merely a thin veil swaying between the appearance of the not-I in the form of a sensible world and the self-activity of the I in the form of thought. Indeed, neither judgment nor understanding play much of a role in Fichte's scheme and his treatment of them seems almost peremptory, as if included only for the sake of "terminological completeness." Of course, it is relatively easy to see why this would be the case, given our prior discussion. Once Fichte has established that the gravamen of the *Wissenschaftslehre* and indeed its driving wheel involves an originary *Tathandlung* and *Grundsatz*, further articulated and made progressively more determinate by other *Handlungen* and *Sätze*, then there is really no further semantic work to be done by the term *Urteil* (or *Urteilskraft*). In Fichte's new configuration, *Satz* serves at least as well as *Urteil* to capture a "propositional" dimension: indeed, in *"Ueber den Begriff . . . ,"* Fichte explicitly distinguishes between the "content" and "form" of a *Satz*, where its "content" would seem to be what is often intended by "proposition" today and its "form" is a function of the particular sort of "speech-act" by which it is articulated.[16] On the other hand, if Fichte wants to refer to an empirical act of judging, then his notion of *Handlung* certainly suffices, with considerable gain in specificity inasmuch as Fichte joins this closely with its articulation in a *Satz*.

IV. SOME REAPPRAISALS

If I have succeeded in making the case for a crucial deep-seated shift in the basic understanding of the relation between *Urteil* and *Satz* as we move from Kant's Critical Philosophy to Fichte's *Wissenschaftslehre*, several reappraisals of prevailing readings of this period and Fichte's relation to it seem in order.

(1) The first thing to note is that this shift, which I have attempted to mark in the writings of Fichte dating from the mid-1790s, seems to have become a permanent feature of the philosophical movement which will follow. Clearly, Fichte's emphasis upon the dynamic relation between *Handlung* and *Satz*, the free activity of consciousness and its objective articulation, becomes the fundamental assumption upon which this tradition will continue to develop that all-important notion of "system." Viewed in this light, Hegel's much-discussed notion of *"der spekulative Satz"* must be regarded as a direct descendent of Fichte's crucial revision of the Kantian conceptual constellation.[17] Indeed, it is fair to say, I think, that, with the exception of a few "doctrinaire" Kantians, *Urteil*

never again appears as a central term of philosophical discourse in the subsequent thought of the German Idealists.

(2) At least a few scholars have attempted to show that the famous "linguistic turn" was actually first made, implicitly if not explicitly, by Hegel rather than Frege, Husserl, Russell, or Wittgenstein (some of the more usual suspects who have been credited with this).[18] However, if one accepts my present reading of Fichte as generally correct, then the point at which this "turn" occurs must clearly be relocated to Fichte's early revisions of Kant's Critical project.

(3) More specifically for Fichte scholarship, this shift from focusing upon Kant's "synthetic acts" and *Urteile* to Fichtean *Handlungen* and *Sätze*, where the latter term is taken to possess an explicitly linguistic dimension, offers a potentially fruitful new line for interpreting the *Grundlage der gesamten Wissenschaftslehre*. In particular, it suggests a way of reading Fichte that may be able to dispose, once and for all, of the well-known Hegelian characterization of Fichte's standpoint as a "subjective Idealism," a reading which Fichte himself was constantly at pains to avoid, even before Hegel had entered the critical fray. It would also assist in helping to square the systematic project outlined in "*Ueber den Begriff . . .*" and the *Grundlage* with the fact that these were almost immediately followed by a long monograph on language.

(4) In the broadest sense, while I do not mean to claim that general issues surrounding the philosophical significance of linguistic matters by any means loomed as large for the German Idealists as they later did for both the Anglo-American and phenomenological traditions, I do believe that there remains much more of philosophical interest—and perhaps even of underlying historical connectedness—than we have yet begun to realize.

NOTES

1. *J. G. Fichte-Gesamtausgabe der Bayerischen Akademie der Wissenschaften*, eds. Reinhard Lauth, Hans Gliwitzky, and Erich Fuchs (Stuttgart–Bad Cannstatt: Frommann-Holzboog, 1964ff), I/4: 184; Daniel Breazeale, ed. and trans. *Introductions to the Wissenschaftslehre and Other Writings* (Indianapolis: Hackett, 1994), p. 4.

2. Cf., for example, Adolf Schurr, *Philosophie als System* (Stuttgart-Bad Canstatt: Friedrich Frommann, 1974).

3. Cf., for example, Fichte, *Gesamtausgabe*, I/ 2: 380ff.

4. *Immanuel Kants Logik: Ein Handbuch zu Vorlesungen*, ed. G. B. Jaesche, in *Kants gesammelte Schriften*, hrsg. von der Königlich Pressischen Akademie der Wissenschaften (Berlin, 1923), IX, p. 101. All translations are my own unless otherwise noted.

5. Here, one might well note that it is precisely this assumed framework that leads to many of the interminable discussions among commentators on Kant who try, though always without success, to remain true both to the "spirit" of Kant's "transcendental" project as well as to the "letter" of the texts in which he attempts to explicate it. My argument here, of course, is that Fichte realized the impossibility of squaring the two and was hence forced to alter some of Kant's basic assumptions in order to remain true to his "transcendental spirit."

6. *Kants gesammelte Schriften*, IX, p. 110.

7. Actually, the situation is more complicated than this, since there he sometimes seems to distinguish between *Grundsätze* and *Principien* as well as between *"oberste Grundsätze"* and mere *"Grundsätze"* (whatever this might mean, given his discussion in the lectures on logic).

8. *Kants gesammelte Schriften*, III, p. 39.

9. Fichte, *Gesamtausgabe*, I/2: 380ff.

10. Fichte, *Gesamtausgabe*, I/2: 148ff. (See also Daniel Breazeale's translation in Daniel Breazeale, ed. and trans., *Fichte: Early Philosophical Writings* [Ithaca, N.Y.: Cornell University Press, 1988], p. 132. In this section, I will quote Breazeale's translation, supplying the appropriate German words where they are crucial to my interpretation of this passage.)

11. It should be noted that Breazeale supplies a "then" that has no warrant in the text.

12. Again, the translation would be more accurate if it read "that which is presupposed," since "presupposition" might be understood as itself a *Satz*, when, in fact, more is actually involved than just this.

13. See Fichte's discussion of the relation between language and freedom in his essay of 1795 entitled, *"Von der Sprachfähigkeit und dem Ursprunge der Sprache,"* in *Gesamtausgabe*, I/3: 98ff. This essay was probably written immediately after the conclusion of his lectures for which the *Grundlage* was the outline.

14. Fichte, *Gesamtausgabe*, I/2: 255ff.

15. Ibid., I/2: 380ff.

16. See, e.g., John Searle's treatment of these matters in *Speech Acts: An Essay in the Philosophy of Language* (Cambridge: Cambridge University Press, 1970), pp. 29ff.

17. For the most complete discussion of Hegel on these issues, see Günter Wohlfart, *Der spekulative Satz: Bemerkungen zum Begriff der Spekulation bei Hegel* (Berlin: de Gruyter, 1981). Wohlfart, however, fails to recognize the debt Hegel owes to Fichte here.

18. The work of Bruno Liebrucks figures most prominently in this respect. The same point has been pursued at length in the recent book by John McCumber, *The Company of Words: Hegel, Language, and Systematic Philosophy* (Evanston: Northwestern University Press, 1993).

9 THE PARADOX OF PRIMARY REFLECTION

Pierre Kerszberg

I

The chief lesson of the whole of Kant's critical enterprise is, perhaps, that we can only try—indeed, we cannot help but try—to speak a posteriori about the a priori. Nothing can be said about the a priori as such. Once one says something about the a priori as such, one makes it an in-itself. Thought must relate to an outside, to the world, without knowing what this outside is in its own being. This is the ultimate meaning of the Copernican revolution that Kant initiated; only by positing an outside can thought "find itself" in itself, but then it never really adheres to itself again. What allows knowledge to reflect upon itself and thus to legitimate itself is that, within knowledge, there is an origin that escapes it. For thought, there is no origin of thought, and yet there clearly must be one. As a result, the legitimacy of any possible knowledge founded on our own ability to think cannot be attached to something other than knowledge itself. This does not mean that transcendental knowledge will eternally contemplate itself. What could legitimate progress, or the movement of knowledge out of itself? We are forced to put the cart before the horse, as it were. An illusion always lies *before* knowledge. This is the necessary illusion of attributing ontological status to knowledge. What is thus ontologized is put *in the place of* the unknowable origin. But could we not locate this origin in the "outside" which thought has expelled from itself? Attempting in this way to recapture the unity between itself and the world, thought engages in the antinomies of the Transcendental Dialectic, which form the very place in which this rupture is reflected in the world.

More generally, the tension involved in this impossible movement of self-foundation in the world outside the mind can be expressed by opposing two kinds of judgment. Let us follow Kant's fundamental distinction between determinative and reflective judgment. The former is the judgment of experience: the universal is given, the particular must be found. In the latter, only the particular is given, and the universal must be found for it. As for teleological judgment, it is reflective in the following sense: it is "the ability we have to subsume the particular laws, which are given, under more universal laws, even though these are not given."[1] In this way, as regards the universal that it seeks, reflective judgment can only but give freely the rule to itself. Suppose that the original source of a concept of the understanding is found in the power of judgment, "it would have to be a concept of things of nature insofar as nature conforms to our power of judgment." This concept is nothing other than the concept of the purposiveness of nature. There is something like a Copernican revolution at work between the power of judgment and the concept of understanding: the concept conforms to the power, instead of the other way around, because without the power of judgment, the understanding could not orient itself in nature. But, in contrast with theoretical knowledge, this reversal has the effect of removing any necessity. If there is anything like a concept that expresses a kind of preestablished harmony between the understanding and nature, this concept would not fall within the purview of the understanding, that is, this concept would not prejudge anything as regards the *contents* of empirical laws. Thus, the more universal laws, which are not given, *could be wholly other* than those of purposiveness. Unlike the category, the concept of purposiveness is *not itself necessary*. By virtue of nature's technic (the concept which arises originally from judgment), even if nature were completely contingent, its laws infinitely diverse and its forms infinitely heterogeneous, we could still produce laws of regularity. This shows that the concept of purposiveness is something forced on nature, a position of authority (Kant says that judgment acts "for its own benefit"[2]) which is no longer necessarily at one with our ability to make do. Kant had begun by arguing that a logical necessity existed in given experience, since a universal (provided by the category) is always already there. But beyond experience, and notwithstanding the logical necessity of the cosmological principle, there is no longer any logical necessity, *not even an illusion* of logical necessity that could be corrected in some way: the concept at the basis of the power of judgment could be entirely different from that of purposiveness.

II

The intellectual journey of Salomon Maimon and Fichte reveals that perhaps even Kant could not remain entirely faithful to his own ultimate principle, which states that "all our knowledge relates, finally, to possible intuitions, for it is through them alone that an object is given."[3] Possible intuitions can be related to a priori concepts in either one of three ways. (1) In mathematics, a pure intuition (such as space or time) is already built into the concept, which can therefore be constructed. (2) In the mathematical principles of pure understanding, the concept contains the rule for the synthesis of possible intuitions which are not given a priori, but which can be represented a posteriori (by means of the rule) in perception. (3) In the dynamical principles of pure understanding, the rule included in the concept enables us only to search empirically for a synthetic unity of that which can never be perceived. Kant also argues that the Transcendental Dialectic results from a "natural" amplification of the dynamical principles referred to as the "Analogies of Experience." This amplification is based on the fact that the pair of mathematical principles ("Axioms of Intuition" and "Anticipations of Perception") gives us a *complete* manifold of intuition, whereas the "Analogies" define only a principle of search because the experience of the manifold is incomplete.

Let us first consider Salomon Maimon, whom Kant himself hailed as one of his most perceptive interpreters.[4] If Maimon were right, it would no longer be certain that the "Anticipations" give a complete manifold to be synthesized in some way. Maimon wants nothing less than to renew completely Kant's notion of intensive magnitude. Kant determines empirical reality a priori with regard to *form* alone, except in the "Anticipations of Perception," where intensive magnitude is a quality defined as the intensity of sensation, or the degree of influence on sense.[5] This degree can always be diminished in thought, but it can never vanish completely. Otherwise, reality would be indistinguishable from its own absence, i.e., negation. Kant explains: "Between reality . . . and negation there is therefore a continuity of many possible intermediate sensations."[6] Consciousness is suspended between givenness and nothingness. Between them, there is realm of possibility that remains pure, in the sense that it never becomes an object of apprehension even though it is the condition of possibility for any given sensation to be apprehended. Thus, it is the *matter* of intuition that is foreclosed by means of the intensive magnitude. At this point, we probably come face to face with the most extraordinary blind spot in the entire *Critique of Pure Reason*.

Maimon reads this argument in the following way: If what is given in sensation was prior to its apprehension in terms of a succession in time, and if this given left no trace in consciousness, then there must have been some kind of conspiracy going on behind the back of sensation.

Maimon ascribes the intractable action of matter to the limited (or unconscious) activity of consciousness. According to him, this activity reveals the presence of a residual quantity that no quality is able to absorb. Concerning the intermediate sensations that remain in the mode of possibility at all times, separating givenness from nothingness, Kant had said that "the difference between any two of [these] is always smaller than the difference between the given sensation and zero or complete negation." Because the "Anticipations of Perception" is one of those transcendental principles of pure understanding proceeding from a concept to an intuition, instead of from an intuition to a concept (as in mathematics),[7] no mention was made of the mathematical concept of differential. Kant has always emphasized the essential difference between philosophical and mathematical knowledge. The former considers the particular only in the universal, while the latter considers the universal in the particular.[8] But Maimon uses the mathematical concept of differential—that is, a quantity that is itself smaller than any assignable quantity—in order to account objectively for the progression, within sensation, from zero to its infinitely small degrees. One possible justification for this conflation between the mathematical and the philosophical method might be the following. When we are doing mathematics, on Kant's own terms of construction of concepts in pure intuition, the form of our synthetical construction may pass off for the matter of that which is so constructed, so that pure mathematics could be regarded as proceeding synthetically in the just the way needed by the a priori concept of matter. As Kant puts it himself, "if pure intuition be wanting, there is nothing in which the matter for synthetic judgments a priori can be given."[9] One consequence of the fact that, in the mathematical synthesis, the form may substitute for the matter of construction, is that there must at the very least be a correspondence between mathematical construction (which takes us outside of the concept) and what Kant calls "the necessity of what belongs to the existence of a thing," even though existence itself is, of course, not constructible.[10]

Maimon, then, needs to find a counterpart for the silent activity of the mind (the infinitely small degrees below the threshold of actual sensation) in the object that presents itself in intuition. He finds that if our sensibility furnishes differentials for a determined consciousness, then the differentials of objects themselves cannot be encountered passively by our consciousness, but are produced by its transcendental imagination. Since the real is now a function of imagination, this solution would lead to an outright skepticism were it not for Maimon's following argument, according to which the unsensed differentials of objects account for nothing less than the things in themselves.[11] Indeed, objectivity would vanish in the absence of any reference to the noumenon, since the jolt provided by external reality would not be needed if consciousness had only the result of unconscious syntheses before itself. Consciousness would never be able to discover a necessary connection of its representations to itself were it not for mathematics, which provides us with the example of such a necessary connection in which the independent content of the appearance is annulled.

This is the intriguing result of Maimon's investigation: in order to be completed, the mathematization of the unconscious moment of sensation requires that the noumenon be rehabilitated. Instead of Kant's two beginnings (the given sensation, on the one hand, and its matter below givenness, on the other hand), Maimon can argue that "our knowledge of things begins at the middle and also ends at the middle."[12] This is what occurs in mathematical knowledge. When we start from a certain magnitude we can proceed either backward or forward by following the same rules of unity, and so we can always think a unit as being larger or smaller than anything given in sensory experience. Arbitrariness has been transposed from the mathematical synthesis to the beginning of experience itself. The way is clear for overcoming Kant's claim that intuitions cannot be reduced to concepts. After Kant, as the history of mathematics has shown, all the logical problems that he thought to be irresolvable as long as this irreducibility was denied, have been resolved by means of the imaginary variable that Kant rejected as part of mathematical science itself.[13] "All of the concepts of mathematics are thought by us and at the same time exhibited as real objects through construction a priori. We are, in this respect, similar to God."[14]

Undoubtedly, Kant's Transcendental Dialectic contains an implicit response to Maimon's development. Kant's theory of transcendental illusion as the unobserved influence of sensibility would enable Kant to resist Maimon's suggestion that the matter of sensation is posited by the activity of the ego. Instead of being an unarticulated accompaniment of the activity of the mind (the barely perceptible signals we receive once we shift into a constructive mood and thus liken ourselves to God), this matter would be ascribed to an excess of appearance that happens not to be fully mastered by the mind, much like the optical illusion that remains even after it has been detected by the appropriate conceptual means.[15] The experience of such a surplus forces human reason to seek the unconditioned *within* the series of conditions of the phenomenal world. But post-Kantian attempts to silence Kant's Transcendental Dialectic did not come only from below (mathematics), but also from above (the transcendental ideal).

III

Indeed, at the other extreme from Maimon we have Fichte, who returns to the ambiguous principle of reflection in transcendental philosophy, which states that a first reflection is both necessary and impossible. He capitalizes on this ambiguity in order to complete the system with respect both to the object and to its subjective conditions. Of course, Fichte begins with the explicit recognition that there is no direct access to the generative act that accounts at once for the emergence of consciousness and for its constituted structure. Actual consciousness has always already outpaced its act of birth, it lies "ahead of" this act. But Fichte thinks that this "ahead" must also have an initial moment. This is the moment of every *new* series in consciousness, which is like "awakening from a deep sleep or from the inability to perceive, especially in the case of a place with which we are unfamiliar."[16] We can sense overtones here of Blaise Pascal's fright at awakening on an unknown desert island. But Fichte is not content to accept the failure of all attempts to reconcile the experience of self-knowledge and the experience of the world. He has great confidence in the new series, which is supposed to allow us to recover the first series in its entirety. Surprised at awakening in an unknown place, I find myself once again in my own being—the Self. Then I turn my attention to the things that surround me, in order to get oriented. This allows me finally to ask legitimately about the *absolutely* first series: Where am I? How did I get there? What has happened to me?

Consequently, Fichte finds that the sought-for amplification of the mind will succeed if we are ready to think thoroughly in accordance with the "duplicity" of consciousness that characterizes thinking, at least while our primary reflection lasts. Is the reflective judgment incapable of giving itself what will be given in the end? It is sufficient to take reflection as a perspective on this end, which is nothing but experience as a whole? Here, the end justifies the means. The abyss between the a priori and the a posteriori is now reduced to two different perspectives on the absolute totality of things. In fact, reflection can afford to be indifferent with respect to what will emerge in the end, because what is given in the *perspective* is also given as part of the absolute totality. Thus, we may ask, do we need two points in order to draw a straight line? In Fichte's opinion, if only a point (the starting point of reflection) is given together with an angle, then "all the points on the line are given along with it" even though experience has not yet run through them.[17] When Kant's dialectical principle of search for new intuitable members of the series toward the unconditioned thus becomes the norm of any possible philosophical activity, the way is clear for rethinking the primordial relation of object to subject. The relation between subject and object can now be considered both *before* and *after* any synthetic activity of consciousness. Before a first synthesis occurs, they are mere opposites, they constitute a pure thought (referred to as the absolute) without the slightest reality. Such a pure thought forms the basis of the first principle of any possible knowledge or being—this is the principle of the absolute self: "The self begins by an absolute positing of its own existence."[18] Kant has taught us that existence is not real predicate. By changing the form of this truth, Fichte gives it a new content as well, i.e., he gives existence its only possible positive expression: Existence is spread *throughout* the judgment by means of the copula that links the subject to the predicate. The first principle, which is a first reflection that does not yet know itself as reflection, can be stated as an affirmation of existence which has no ground that can be thought or stated in the form of a judgment. Clearly, in order to posit the first principle, Fichte needs to violate Kant's rule for any possible transcendental proof: Instead of viewing the reciprocal movement of the ground and what is grounded as one that cannot be overcome,[19] Fichte says that the ground must fall outside what it grounds precisely because the former is supposed to explain the latter.[20]

The positing of the absolute self is merely the coming into being for us of what is formally identical to itself in its own being. But precisely because it is absolute, this form is still nothing determinate for us, even though our consciousness has apprehended it. If we then reflect upon the conditions that have enabled us to posit the first principle, we will oppose a second principle to the first. The second principle thus finds its expression in the form of a contradiction: a not-self is opposed absolutely to the self.[21] Again, nothing is thereby determined for us, because the not-self, too, is posited absolutely. If the two absolute principles are combined absolutely, they will simply annihilate each other, which is another way of proving that they are still nothing for us. The next reflection focuses on what allows these two opposites to coexist in our own consciousness. This gives rise to the third principle, according to which the self posits itself as limited, and thus *determined* by the not-self. A synthesis of the two principles is thus possible if, instead of mutually destroying each other, they limit each other, that is, if they are divisible so that a certain quantum of self coexists with a certain quantum of not-self. Fichte achieves what was impossible for Kant. Because the movement from one reflection to the next takes place in an absolute totality that is always already given, it gives us an insight into the actual genesis of representation. Fichte characterizes such a process as a *deduction* of what is presupposed in any determination. At the beginning, our consciousness is filled by nothing, except by what Fichte calls a "salutary illusion of the imagination which, without my knowledge, ascribed a substrate to these pure opposites."[22] The transcendental illusion is thus no longer as irreducible as it was for Kant's notion of reason. To be sure, we cannot begin by reflecting upon nothing, but our deceptive imagination provides us with the elementary material that was needed. Fichte finds that the illusion itself was not to be avoided or eliminated. All we had to eliminate in order for reflection to materialize itself into determination was the product of the illusion. The absolute totality is no longer an Idea of Reason, but something imagined, which Fichte conceives, in the final analysis, as allowing us to deduce the sum total of categories and principles of understanding.

On balance, Fichte attacks transcendental philosophy from above, while Maimon attacks it from below, and both aim at surpassing it. Maimon discovers that sensation would now be given immediately by the thing in itself, so that the transcendental possibility of its synthesis—and not just its final product—would already require mathematics. If *all* degrees from 0 to *x* can be anticipated only by including the thing in itself as the matter of sensation, Kant would have come close to contradicting himself when he said, as a general truth of critical philosophy, that "the matter of appearances, by which *things* are given us in space and time, can only be represented in perception, and therefore a posteriori."[23] Perhaps, without Kant being able to realize it, the "Anticipations" have already brought us into the dialectical domain of necessary illusion. If it were the case that the "Anticipations" provide us with an incomplete experience, an enlargement of the scope of Kant's Dialectic would be made possible; we would have more than three Ideas of Reason. This prospect justifies Fichte's strategy. The whole world is within us. The world and the concept of the world are but two perspectives on the same thing. To reach this point, Fichte borrows his model from that of God's thinking: "Our idea of

God and God are the same thing, and this idea is objective, that is, it is something from which we cannot separate ourselves and which is as certainly in us as it is certain that we exist."[24]

IV

Does Kant's own Transcendental Dialectic give us any clue as to how it might be enlarged? Indeed it does. We can find this clue if we follow the direction given in Fichte's *Science of Knowledge*. Fichte completes Kant's project by reversing the orders of truth and illusion, that is, by taking *explicitly* the Dialectic as the core of critical philosophy. The absolute of the first principle is already a necessary illusion, because it can be posited without being anything for us. In other words, Fichte's method might indicate how to exhaust the Dialectic; since the Dialectic is its starting point, in principle, all its consequences can be drawn from it. In fact, the theoretical part of the first *Science of Knowledge* (1794) concludes with the assertion that a reciprocal action exists between the self and the not-self. Neither has the exclusive privilege of being finite or infinite. On the contrary, once either term is infinite, the other is finite. According to Fichte, this is the basis of the antinomies Kant established.[25]

We are brought back to the ultimate horizon of the whole critical enterprise. In the very last lines of his *Opus Postumum*, Kant seems to exalt the poetic and mythic figure of Zoroaster as the sole prototype representing the highest end of human life: "Zoroaster: the ideal of physical, and, at the same time, moral-practical reason unified in a single sensible object." For humans in general, philosophy is no more than a perpetually unfulfilled "effort toward [such] wisdom."[26] But might not this effort be reduced to nothing in light of the ultimate ends?

Now, for Kant, any imagination that supposedly intuits the future is simply deluded.[27] Far from allowing us to speculate freely about possible future events (as was the case for David Hume), the imagination only suspends the present course of events. It makes us reverse the order of time: either we invert the priority of experience over our capacity to understand, or, retaining this priority, we project the prototype of a "primary experience" far into the past, back to an absolute origin that we presume contains the cause of *all* the phenomenality of experience. From the transcendental point of view, the reversal is always double. Through a mode of consequentiality proper to the categories, the priority of the understanding, which conditions all phenomena, inevitably leads us to posit something absolutely unconditioned that encompasses all conditioned phenomena. In other words, the only way to overcome Hume's skepticism is to prove that the future is not at all a transcendental problem, something that Kant does in the cosmological antinomies of pure reason, where he shows that the unconditioned is either the ultimate point in the past or the total series of past events.

Of course, the future would become a problem for practical reason, because the notion of the sovereign good demands that the reign of nature and the reign of ends be ultimately reconciled. But were we immediately to move on to consider the practical realm, we might leave something unfinished in the consideration of the theoretical. The same circumspection that led us to delay converting the idea of totality into an ontological problem is in order here. Before moving to the practical, we might ask why the future is not even an inevitable illusion of theoretical reason. For Kant, reason does not suppose the question of the totality of consequences of possible conditions, for consequences do not make their conditions possible.[28] Consequences are an arbitrary and not a necessary problem for pure reason.

It may be apt to recall that, as Heidegger has pointed out, the rejection of consequences as a rational problem is valid for corporeal nature only, and certainly not for history. For a historical event, understanding it by means of its consequences is essential. This is true for the event at any time: even a past event must be determined by the possibility of its future—this future could be our own present. That is why, Heidegger goes on to argue, "the history of the present is a non-sense."[29] Are we then committed to Heidegger's view, according to which Kant simply ignores this other dimension of being, as if the Kantian field of phenomena were exhausted by the domain of the ready-to-hand? It cannot be denied that Kant might well have good reasons for rejecting the consequences as a rational problem, but he certainly does not fail to take them into consideration: He says explicitly that there is no Copernican revolution in history, because this would require a viewpoint that can both see and predict the free actions of human beings.[30] Unlike the Copernican revolution in astronomy, where a position of rest within one reference frame is always possible, there is no such viewpoint for us in the historical course of human events; we are always sent back to another reference.

History does not justify freedom. But could we not say, conversely, that freedom justifies the very movement of history? Fichte tried to develop such an approach of history in his early reflexions on the French Revolution

(1793). In the immediately subsequent *Science of Knowledge*, he goes on to give a theoretical basis to this possibility. In order to understand this, let us go back one more time to Kant's sense of arbitrariness concerning the future.

There seems to be a correspondence between the division of the problem of pure reason into reality (the past) and chimera (the future) and a certain restriction of the problem of reason in Kant. Reason does not produce any concepts; it only frees the already existent concepts of the understanding from the limits imposed by possible experience. It is a matter of achieving the complete comprehensibility of what is given. This implies the given reality of the regressive synthesis, which points toward ever more remote conditions in the past. The progressive synthesis, on the other hand, is only *capable of being given*. The given contains within it a possible totality that can always be unfolded. This means that the present is pregnant with the past since it is already by itself a synthesis. To unfold that which is contained in the synthesis is to retrieve something forgotten: one runs through the synthesis contained in the given in an opposite direction to the one which occasioned it. Thus, what a synthesis is by itself is, for us, to be unfolded as an antithesis.

Fichte gives a purely formal justification for the inseparability of synthesis and antithesis, outside the temporality Kant assigns to regressive synthesis: "There can be no antithesis without a synthesis; for antithesis consists merely in seeking out the point of opposition between things that are alike; but these like things would not be alike if they had not first been equated by in an act of synthesis."[31] Retrieving what has been forgotten does not pull us back toward the past. And yet we are still within the realm of transcendental logic. That is why the form in question here is a very peculiar one. If ordinary concepts are set against one another, we can raise ourselves to a higher concept (one with greater extension) that combines them in a synthesis. Conversely, two identical things seem opposites from the point of view of a lower concept to which one can descend. The self, however, as something merely posited, must contain the absolute totality of all conditions. A not-self can thus be opposed to it only formally. The fact that the two terms are brought together in one sphere does not raise us up to a higher concept, since one of them is already absolutely comprehensive. To reach a synthesis, we have to conceive a series in which the absolute self moves toward a lower concept. To the absolute self, Fichte thus attributes the character of divisibility, one which is more spatial than temporal.[32]

Fichte's general criticism of Kant is that he limited the scope of the transcendental deduction of the categories. Kant's task in the deduction was simply to show that *there are* categories, and not which categories belong specifically to the human mind. Similarly, crossing the boundaries of the understanding, why does reason limit itself to just these concepts? The Idea of Reason seeks to extend the concept of the understanding beyond the bounds of the empirical while retaining a connection with it. At a certain point in the past, when we raise ourselves toward higher conditions, the connection with the empirical is broken for us, though we presume that this connection must continue to exist in itself. Regarding the descent toward consequences, could we not say that the present constitutes the same kind of rupture? Future events would thus constitute a leap into the in-itself, justified by the same concern with continuity. In the regressive synthesis, one cannot necessarily see *at what point* the rupture occurs in the continual slippage of the phenomenal part of the series into its transcendent part. There seems to be just as much arbitrariness in the series of the past (knowing at what point the rupture occurs) as there is in that of the future (knowing whether the series will come to an end). One may transpose the present into the past in order to make the latter seem given, but only to a certain extent, for this transposition is an exercise in pure thinking. Kant says that transcendental illusion results from an unobserved influence of sensibility on the understanding. This means that, without my being aware of it, the present forces me to think of the past as a series of present moments that are identical to one another. If this influence were perceived, the sensible in all its diversity would regain its preeminence. To think of the past series from the standpoint of the present is thus a purely formal exercise which makes all moments homogeneous. Because the present is the only event that is of itself (without the intervention of thought) at once a form and a content, it is just as much an ultimate condition of possibility as the unconditioned sought in the regressive synthesis.

Fichte understood the eminent role of the present in the antithetic. The present pushes the antithetic into another order of time. He writes: "There is no first moment of consciousness; there is only ever a second moment."[33] Our reflection on what has happened in the past must be distinguished from the past that necessarily exists for us. In itself, the past is not. It exists only for a consciousness that is possible in the present, one that reflects on the condition of the present and wonders whether some time has actually passed. Since the past truly exists only in an inquisitive reflection, reflection is not hindered in its freedom by the activity of the prior series. The present is far from being completely weighted down by the past. Rather, Fichte writes, "in it, any other perception could just as well have arisen." The regression toward the past unconditioned retrospectively runs through perceptions that could not have been different than they were. It is made possible by consciousness in its *identity*. This process of retrospection can go on indefinitely, because all differences have been eliminated. But the *freedom* of consciousness is not limited to this identity. Consciousness is still free to direct its reflection to other objects besides the past, in order to recover difference.

V

Though for Kant any imagination that intuits the future is simply groundless, the passage from the order of determination to that of *reflection* leads Kant to rediscover the future as a transcendental problem. This, indeed, is the absolute fragility of the principle of reflection. On the one hand, since the power of judgment must presuppose itself as regards the universal that it seeks, it cannot but project itself onto experience. On the other hand, in order for "judgment [to] subsume the particular under what is universal though still always empirical"[34] (because the universal is still sought), experience needs the systematic organization of its empirical laws. We run up against a radical impossibility of establishing a first foundation: the founding act, as it were, always chases after itself in reflection. The power of judgment puts itself to work only when it is already at work; it is in its activity that it recognizes itself. Thus, it is not the lawfulness of judgment alone that accounts for the lawfulness of purposiveness. It is as though the freedom that nature chooses for itself corresponded to the purposiveness we put in nature. But this correspondence remains contingent, and so the principle that reflection chooses for itself cannot quite escape the threat of arbitrariness. Reflection would be traversed by an antinomical conflict of the type of rational cosmology—as Hegel in fact claimed—if the power of judgment necessarily had to assume that the purposiveness that it projects was realized objectively. But the power of judgment *does not need* to suppose nonpurposiveness in order to assert purposiveness. This power is endowed with an absolutely unconditioned freedom to posit anything it likes (this is its heautonomy), so that in the antinomy of teleological judgment it can limit itself by claiming (in the antithesis) that only some products of nature must be explained in accordance with purposiveness. It cannot limit itself by positing something other than itself, because the mechanical explanation (in the thesis) cannot even be denied.

Kant's reflective judgment makes itself impervious to error by claiming to have in its grasp an unconditioned that it has not found outside itself. In this way, despite its absolute fragility, reflection does not infringe upon the form of determination. Only the future of reflection (history and its vicissitudes) will show whether the concept of purposiveness was necessary. However, it can hardly be denied that by giving itself the law now, the faculty of reflective judgment dogmatically extends the present to all future eternity. Paradoxically, its assertion of freedom (through heautonomy) cripples freedom for all time to come. Fichte brings reflective judgment back into the sphere of error and truth through the intermediary of what he calls an obscure feeling.[35] Instead of considering freedom (in heautonomy) to be a nascent reflection that is always reiterable—because any guarantee for the corresponding universal will always be missing—we can give the superhuman unconditioned a real weight. This weight so overwhelms reflection that, in the extraordinary profusion of laws, reflection can no longer find itself. Lost in a morass of rules, judgment cannot avoid confusing one rule with another.[36] It must therefore descend into the series of conditions and lose itself in the play of positing and counterpositing. Instead of placing the superhuman in the human as evidence of the unreflective and insatiable desire of reason, Fichte asks us to engage in a deliberate ruse with the infinite to facilitate its entry. We should then be able to find the key to the initial problem of all transcendental philosophy, which consists in the unprepared irruption of the infinite as the framework of all finite experience.

We have to reflect in such a way that that which exceeds us (the not-self) does not exist, though it is still there.[37] This regulated reflection is the work of the imagination, by means of which a repetition (*Wiederholung*) of the self's act of self-positing opens it up to an exterior influence, a check (*Anstoss*) initiated by nothing determinate.[38] If one can distinguish the activity of the self from its product, the repetition of the same can produce something new. One can then receive an x that was found in no past synthesis or antithesis. How, then, are we to reconcile, on the one hand, the encounter with this check, and on the other hand, an act of understanding that lies entirely in the self? Fichte undertakes something like a bracketing of the Copernican revolution (in Kant's sense). Since the understanding and the encounter are not identical activities, the question still remains "which of the two is to align itself to the other."[39] The answer is clear. Everything must depend on the absolute self. Since the absolute self posits itself, in the relation to a possible object, it agrees only with the *form* of this object, and not its actual identity. In other words, in its relation to the object, the self discovers itself to be an infinite effort, tending towards mastery over the form. This discovery is, however, a result of the bracketing and not its condition of possibility.

Of course, effort (or tendency) has no causality; it is not primordially determined by any object, but we can reflect on it *as if* it had a cause. Indeed, by virtue of reflection, the self at least exercises some "causality upon itself."[40] Instead of claiming that reflection is not directed toward anything determinate, one might just as well say that in its absolute freedom it does have to direct itself toward something, even if only unconsciously. The direction is provided by conformity to duty.[41] Tendency does not produce an object, but this conformity does explain *feeling* as a determinate tendency, a push toward an unknown, which is nonetheless something.[42] While Kant refuses to oppose nonpurposiveness and purposiveness, Fichte opposes activity and nonactivity in terms of consciousness and unconsciousness. The self does not see the production of the object it produces; this production is unconscious. The self is thus led to deny its own activity temporarily.[43] The activity is restored by means of another act (a new reflection) which has the peculiar property of interrupting the act of determining the x that is foreign to the self.[44] This new doubling of the self's positing has the effect of giving the form of the x, which is produced (unconsciously) by the self, the character of an *image* of this x. Clearly, there are several images possible for each form, so that the product of the self "is posited as *contingent*, as something that did not necessarily have to be as it is."[45] One moves from the form to the image of the form in the same way one moves from feeling, which is related to the real without seeing it, to intuition, which does see, though ineffectually. How to make feeling see? The self could not even live if it had no feeling of anything. But how can it direct itself toward something determinate if it is an irresistible push towards the unknown? In response to this question, we will have to say that the interruption of the act of determination indicates another tendency in the self, "the drive towards change of feelings," which Fichte defines as the object of *longing*.[46] Though it is no more determinate than feeling was, the change of feeling at least removes the blindness from feeling. Instead of not seeing at all, it can see without being able to account for the act of seeing. Finally, it can legitimately pretend to see, for it does so in accordance with the self's freedom: "the other feeling that is longed for, the self can determine it in itself as it pleases."[47] That which is different from the unknown remains equally unknown, but the self can operate as it pleases between the two with the satisfaction of not violating the rights of the final object it infinitely aims at without ever knowing this object. The arbitrariness of our freedom no longer annihilates the higher freedom of the absolute, nor is it annihilated by it. Feeling is now accompanied by approval. Through the yearning that I experience, I also experience the feeling of being drawn up by a world projection in which all the objects of tendency converge. This project does not exist, for it has moved ahead of me into the warp and woof of the world. Yet, it is there, since it draws me up and constantly pushes me into a future that is not just anything at all.

The whole deduction of feeling and intuition has taken place as if in a dream. Of course, the dream is a traditional sign of the future. In it, I am possessed by a meaning which escapes me. But even while dreaming, I feel that if the dream were interrupted, I would become the master of meaning once again.

In this way, Fichte seems to have overcome what, in the Transcendental Dialectic of the first *Critique*, Kant called the vice of inverted reason. Reason does indeed attempt to determine what the understanding, in its empirical employment, is unable to determine. By means of regressive synthesis, it proceeds toward ever more remote conditions in the past. But to begin from an origin amounts to hypostatizing the reality of the principle of final unity. This would mean that we are dictating ends to nature.[48] Reversed reason can thus be called a vice in the moral sense of the term, for in the moral order it is forbidden to base obligation on a source outside oneself. But Fichte finds a way to reconcile absolute arbitrariness and absolute necessity. What sense does this have for our lives?

In order to discover the first principle, one may begin, Fichte tells us, with an arbitrary fact of consciousness, since consciousness is not yet anything determinate. But what exactly is a primary reflection that is guided by nothing? We must at least have a presentiment of what will come of it. Since we already know that our decision will ultimately be grounded in a higher necessity, we must begin by admitting that the arbitrary point at which we begin is so only superficially. Thus, Fichte in reality begins from one layer of the world "below" the fact, one that stands beneath the immediate appearance of appearance. It is not given, and yet it is something that is there, which guides the first contact with the surface. This something that is there is existence, life, though not as an idea, but rather in its very flesh.[49] The absolute self is like the body. Inside there is a given that exists, but which is not really there for me. I experience it most often through pain, or just the absence of satisfaction. On the surface, however, there is the skin; something that is there, but which one can remove, turn over or distort almost completely arbitrarily. Let us call this "there," which is never given once and for all, the *texture* of my world. One can then understand why, correlatively, in the *Science of Knowledge* the ultimate horizon of reflection is yearning. The self first operates as it wishes; it can posit an absolute which, as a world texture, lends itself to all manner of manipulation. Yet, the arbitrariness of the movements on the surface of the body disappears, and leaves room to the inescapable, once one penetrates more deeply into it. Similarly, the self acts from the outset (though without knowing it, i.e. by means of imagination alone) in conformity with the real givens of the world, which exceed its private sphere: the ultimate (unknown and unknowable) ends of rational freedom. When the self opens up to the Other, it only discovers what is always already given in itself without being there. It is therefore the absolute totality, which contains at once everything in it and outside it.

NOTES

1. I. Kant, in "First Introduction," *Critique of Judgment*, trans. W. Pluhar (Indianapolis: Hackett, 1987), p. 202.

2. Ibid., p. 203.

3. I. Kant, *Critique of Pure Reason*, trans. N. K. Smith (New York: Macmillan, 1929), A719/B747.

4. See Kant's letter to Marcus Herz, May 26, 1789, in *Kant: Philosophical Correspondence*, trans. A. Zweig (Chicago: University of Chicago Press, 1967), p. 151.

5. Kant, *Critique of Pure Reason*, A166/B208.

6. Ibid., A168/B210.

7. Ibid., A160/B199.

8. Ibid., A714/B742.

9. I. Kant, *Prolegomena to any Future Metaphysics*, trans. J. Ellington (Indianapolis: Hackett, 1985), p. 27.

10. I. Kant, *Metaphysical Foundations of Natural Science*, trans. J. Ellington (Indianapolis: Hackett, 1985), p. 6.

11. S. Maimon, *Versuch über die Transzendentalphilosophie* (Berlin: 1790), in *Gesammelte Werke*, ed. V. Verra (Hildesheim: G. Olms, 1965–1976), vol. 2, p. 50.

12. Ibid., p. 350.

13. See J. Vuillemin, *Physique et Métaphysique Kantiennes* (Paris: Presses Universitaires de France, 1955), p. 52.

14. S. Maimon, *Über die Progressen der Philosophie* (Berlin: 1793), p. 20.

15. Kant, *Critique of Pure Reason*, A297-8/B354.

16. *Grundriß der Eigenthumlichen der Wissenschaftslehre in Ricksicht auf der theo-*

retischen Vermügen, in *Johann Gottlieb Fichtes Sämmtliche Werke*, ed. I. H. Fichte (Berlin: Viet & Co., 1845–46), I, p. 362. Reprinted, along with *Johann Gottlieb Fichtes nachgelassene Werke* (Bonn: Adolphus-Marcus, 1834–35), as *Fichtes Werke* (Berlin: de Gruyter, 1971).

17. *Fichte: The Science of Knowledge*, trans. John Lachs and Peter Heath (New York: Appleton-Century-Crofts, 1970), First Introduction, p. 26.

18. Ibid., p. 99.

19. Kant, *Critique of Pure Reason*, A737/B765.

20. *Fichte: The Science of Knowledge*, First Introduction, p. 8.

21. Ibid., p. 104.

22. Ibid., p. 200 (translation modified).

23. Kant, *Critique of Pure Reason*, A720/B748.

24. *Nachgelassene Schriften*, bd. 2, p. 299.

25. *Fichte: The Science of Knowledge*, p. 217.

26. I. Kant, *Opus Postumum*, Akademie Ausgabe, vol. 21, pp. 4, 6.

27. Kant, *Critique of Pure Reason*, A222/B270.

28. Ibid., A331/B388, A336–37/B393–94, A410–11/B437–38.

29. M. Heidegger, *Vom Wesen der menschlichen Freiheit* (Frankfurt a.M.: Klostermann, 1982), *Gesamtausgabe*, vol. 31, p. 213.

30. I. Kant, *The Conflict of the Faculties*, Akademie Ausgabe, vol. 7, pp. 83–84.

31. *Fichte: The Science of Knowledge*, p. 112.

32. Ibid., p. 116.

33. *Grundriß der Eigenthumlichen der Wissenschaftslehre in Ricksicht auf der theoretischen Vermügen*, p. 410.

34. Kant, "First Introduction," *Critique of Judgment*, p. 203.

35. J. G. Fichte, *Ueber den Begriffe der Wissenschaftslehre*, (1794), in Fichte, *Gesamtausgabe*, I, p. 143.

36. Ibid., p. 146.

37. *Fichte: The Science of Knowledge*, p. 189.

38. Ibid., p. 243.

39. Ibid., p. 230.

40. Ibid., p. 258.

41. Ibid., p. 259.

42. Ibid., p. 261.

43. Ibid., p. 262.

44. Ibid., p. 276.

45. Ibid., p. 277.

46. Ibid., p. 280.

47. Ibid., p. 283.

48. Kant, *Critique of Pure Reason*, A693/B721.

49. This analogy between Fichte's absolute self and my own body is suggested by what Merleau-Ponty has to say about the "corps propre" in *La Structure du Comportement* (Paris: Presses Universitaires de France, 1942), pp. 230–31.

SCHELLING'S
VOM ICH ALS PRINCIP DER PHILOSOPHIE AS A READING OF FICHTE'S *GRUNDLAGE DER GESAMMTEN WISSENSCHAFTSLEHRE*

Michael G. Vater

I

Fichte wrote the *Grundlage* in great haste in 1794–95, though he first formed the idea of a philosophic system built upon the I in 1791.[1] Before claiming his professorship in Jena as Reinhold's successor, Fichte had crafted a prospectus of the new philosophy, meant to attract students to his lectures.[2] In "On the Concept of Theory of Science," he promised a system that not only satisfied Reinhold's formal demand—philosophy must be based on universally admitted principles—but adopted Reinhold's ultimate fact—the subject-object structure of consciousness—as its content. Fichte planned the *Grundlage* as a course book to accompany the first set of lectures; it was written, printed, and distributed in installments to Fichte's students and selected friends. It was never meant for public view as either a popular or technical statement of the system of transcendental philosophy.

Someone unfamiliar with Fichte's systematic intentions, as announced in the "Review of Aenesidemus" and part 3 of the prospectus essay, would have found reading the first number of the *Grundlage* a demanding task.[3] Fichte had a vast capacity for sustained and detailed argument, but he rarely stepped back to a wider framework to provide transitions, overviews, or simple statements of the conclusions that his arguments advanced. Even if by August 1795 the attentive reader had the whole work before her and could appreciate the practical part—which employs a novel psychological vocabulary to construct the subconscious (or in principle unconscious) platforms for modeling empirical consciousness—there were still few clues about the author's systematic intentions and how these particular discussions of presentation and feeling advanced them.

Before Fichte came on the scene in 1793–94, Reinhold had cleared the ground for a system of transcendental philosophy by demanding that Kant's writings be turned into a philosophical system. Yet his own work substituted for a *philosophical* system a popularization of Kantian epistemology, for Reinhold could not think any farther than the sheer givenness of subject–object polarity in empirical consciousness.[4] Fichte takes the same contents in the *Grundlage* and goes beyond the "facts of consciousness" to a foundational or "principled" deduction of the being-for-a-subject of objectivity as such (i.e., a deduction of presentation) and of the subject's drive-to-alter-objectivity as such (i.e., a deduction of appetition).[5] It is these deductions that transform facts of consciousness into Theory of Science. Fichte added argument or logical rigor to support the "facts" and so turned the Kantian transcendental (or heuristic) analysis of consciousness into theory, or as it was then said: science.

Fichte publicly laid claim to this accomplishment in the "Introductions" to Theory of Science he published in 1797–98. What he does not do is explain the peculiarity of this first version of the first *Wissenschaftslehre* and its tortured deductions. I find it is similar to Gottfried Leibniz's *Monadology*: the construction of "spiritual substance" as a psychic machine driven by the opposed forces of perception and appetition.[6] Like Leibniz's elegant metaphysical construction, Fichte's deduction of objectivity or empirical limitation inside consciousness has two interrelated sides: what from the cognitive side supplies objectivity because it is felt to be sheer limitation or "*check*" is from the practical side self-affection or the non-causing causality of *striving*.[7] Though this double deduction of objectivity (i.e., the set of necessary conditions for empirical consciousness) is in its own right an argumentative tour de force, the basic task of the Theory of Science is to show transcendental idealism: explanation from the point of view of the experiencing subject, free of contradiction. As Fichte read Kant's text, Kant's philosophy was not free of contradiction, especially in its unargued adopted of the "thing in itself" as the ground of objectivity. In place of this ad hoc (or unexplained) explainer, Fichte's deduction of objectivity provides a coherent platform for anchoring more detailed accounts of logic, knowledge, nature, society, law, and morality and for theoretically unifying them all as products of the I's self-realizing activity or spontaneity. Fichte himself seems unaware that "system" is possible only as a coherentist, not a foundational program.

Fichte does not announce that the *Grundlage* works a highly abstract abstractive reflection upon the I's activity,[8] and that it uses arguments both intricate and pedestrian to convey *intellectual intuition* of the I in its completely skeletal, transcendental (i.e., wholly nonempirical) shape. Nor does he admit that the work provides only foundations for an eventual system, or better, a logical canon for all possible systems that do not in principle exclude an account of consciousness. Some early programmatic statements by Fichte suggested that an idealistic philosophy as a totality would connect empirical cognition with action and resolve the object-dependence of cognition into the infinite moral task of object-conquest. Though Goethe's stage manager might promise scenic excursions through heaven, earth, and hell, Fichte makes no such extravagant promise in 1794–95.[9] He cannot at the start display the whole pageant of the realm of consciousness: sensation, matter, nature, individual will, community, world, and providentially ordered history. Schelling will do this concisely and beautifully in the 1800 *System of Transcendental Idealism*, a work that lives up to its name. Eschewing this large canvas and grand theme, Fichte first tackles the question of the foundation of idealism at its problematic core: *if* there is to be *idealism,* one must find an explanation for objectivity and for the object-dependent states of presentation—which run through and mediate all acts of empirical consciousness, volitional and affective and well as directly cognitive. If at any point objectivity is explained by objects, by "things" on which the I and its activity depend, idealism is abolished and the freedom, spontaneity, and self-positing activity that idealism seeks to defend are swept away.

For one wishing to make a philosophy of Kant's Criticism, Kant's resort to a "thing in itself" as a final ground of reality and objectivity was more than a minor difficulty: Some *account* of objectivity is needed, some explanation of the intractable resistance of the known to alteration by consciousness and for the imperviousness of empirical reality to alteration by will. Absent this, presentation would be indistinguishable from dream, present sensation from one imaginatively reproduced. But if the philosopher takes the realistic path and ontologically privileges objectivity, he makes knowledge a commerce of things imaged and things "without" and the knower becomes a machine among things, a shuttle shifting between woof and warp, not the activity of relating, interrelating, self-relating. Realism can product "picture theories," but *never* a viewer of the picture. There is no inching into realism, no quiet accommodation with dogmatism. Reinhold had made all the accommodations; battling for a textually "correct" Kant, he had lost the war for idealism. The activity that is I and does I must be beginning, middle, and end for transcendental idealism.

II

In 1795 Schelling was finishing his theological studies at Tübingen, where he had deeply studied Plato and Kant. Irritated by the "theologizing" Kantians there who wished to use Kant's moral postulate of God's existence to make quick work of their apologetics, Schelling made the question of the possibility of a systematic transcendental philosophy his own.[10] Though he has been represented as a mere popularizer and disciple of Fichte's early in his career (not least of all by Fichte himself)[11] Schelling in many ways shows himself to be Fichte's equal in the years of supposed "discipleship." In 1794–1797 Schelling is more consistently interested in the scope and completeness of systematic philosophy than Fichte is, while Fichte is more careful about guarding the transcendental perspective and securing its foundations. Schelling's taste for abstraction pulls him away from the transcendental perspective, both in theoretical philosophy and in practical domains such as ethics and philosophy of history. In the early essays that Fichte was pleased to read as evidence of discipleship I find more metaphysical anticipations of the identity philosophy of the 1801 *Presentation of My System* than I do evidence of a careful thinking along with Fichte. The latter's detailed phenomenology of cognition and volition is missing; in its place is the metaphysical scaffolding for the grand architecture of system.

Though both philosophers use some version of the contrast between dogmatism and criticism to situate their vies, Schelling is consistently attracted (and repelled) by the explanatory seamlessness of dogmatism, personified in the steel rigor of Baruch de Spinoza's axiomatized metaphysics.[12] Though he sometimes allies himself with a pure transcendental position from 1794 through 1800, Schelling is receptive toward Spinoza's fatalism or the absence of freedom, at least on the level of empirical volition.[13] Or to put it another way, Schelling lacks Fichte's vivid intuition that spontaneous activity is the core of selfhood, or that the I is self-realizing as self-thinking. He prefers the third-person grammar of *production* to describe the transcendental subject and its activity, while Fichte favors the first-person language of *self-positing*. In *On the I as Principle*, Schelling uses the terms "I" and "the absolute" interchangeably.

As I read Schelling's essays of 1794 and early 1795, I find Spinoza as obvious an influence as Fichte. Accordingly, I find little surprising in the metaphysical ambivalence Schelling voices later in 1795 in the *Philosophical Letters on Dogmatism and Criticism.* Though he there ranks the *Theory of Science* alongside Kant's *Critique* as a universal standard to measure all possible philosophies, he nonetheless finds naturalistic metaphysics equally choiceworthy as philosophy of freedom.[14] All theoretical philosophies have posed the same unanswerable problem: Why is there experience at all? (Kant); Why has idea stepped out of the absolute and become opposed to objects? (Spinoza, rephrased by Lessing); Why are my perceptions accompanied by the feeling of necessity and unalterability? (Fichte). Schelling finds the basis for choice between systems is personal and idiosyncratic: logically considered, the option for fatalistic self-annihilation under dogmatism is as cogent as is the choice for autonomy.[15]

We now turn directly to Schelling's *On the I.* The essay was occasioned, inspired if you will, by receipt of the first fascicle of the *Grundlage.* Its title reflects that inspiration, and the first eight sections paraphrase of Fichte opening three sections on the fundamental principles, both in their content and their vocabulary. That the reflection is direct can be seen in Fichte's reception of it as a popularization of his own work. Comments he made to Reinhold about the his dissatisfaction with the literary form of the *Grundlage,* and about the desirability of a reader linking up with his intuitions, not his words, show Fichte was more interested in having others share his general position than he was in their recitation of a catechism.[16] Fichte, however, seems not to have noticed that Schelling's adherence to the transcendental position soon wears thin in *On the I,* just as it had in Schelling's first work, *On the Possibility of a Universal Form of Philosophy.* The public noted their difference more carefully; in his historical review, Reinhold suggests that Fichte and Schelling made the breakthrough to a purified Kantian philosophy at roughly the same time.[17]

How faithful a reflection of Fichte's line of thought is found in even these opening sections of Schelling's essay? Fichte's style of thought is original and rigorously systematic or deductive; his writing is generally a long march from hypothesis to conclusion, uninterrupted by metacomment or historical comparisons. Schelling, on the other hand, is a synthetic or historical thinker who works at some distance from direct hypothesis and argumentation, though he will argue to cinch a point. When Fichte speaks of a philosophy founded on principles and of the necessity for an unconditional principle, he seems to be making a plainly logical demand. When Schelling paraphrases the same arguments one sees—as in the most

Fichtean parts of the 1794 *Universal Form* essay, where absolute philosophy is viewed as an interpenetration of form and content, and where *what* is to be thought determines *how* it is to be thought—that the discussion is also driven by historical figures and their similar styles of argument, e.g., by Plato's quest for a nonhypothetical and deductive science or by Reinhold's search for a philosophy secured by universal principles. In that essay Schelling in fact takes Fichte's I—the principle that unites form (identity) and content (selfhood)—as but one convenient illustration of this deductive model of absolute philosophy.[18] Since his concern is more with metaphysics than epistemology or psychology, he feels free to abstract the logical content from Fichte's three basic principles and use the so-called laws of identity, sufficient reason, and synthesis to generate Kant's quite unexplained table of categories.[19]

Schelling's attempt to deduce Kant's categories in *Universal Form* is an original effort on his part to unify Kantian philosophy, as is the final section of *On the I* where he brings all the forms of judgment Kant discussed under the general heading of a modal synthesis which progresses from possibility, to actuality, to necessity. By contrast, Fichte uses Kant's categories in the theoretical section of the *Grundlage* (§ 4) in a "destructive" rather than deductive manner: the argument reduces all the categories of relation—cause and effect, substance and accident, and reciprocal determination—to the paradoxical idea of a "determinate determinability." When thought gives up trying to think this thought and "imagination" is brought in to reinterpret it as the wavering inside and beyond a boundary that is intuition, and when that interpretation is surpassed as well in the curious alienation of productive activity to a "fictive" not-I in the "Deduction of Presentation," it seems that Fichte has dissolved the theoretical into *aporia* and that only recourse to models of *action* will permit the stabilization of any discourse about cognition.

Even when Schelling is conceptually the closest to Fichte in *On the I*, he speaks a different language. In the first section of the *Grundlage* Fichte describes the I as pure self-positing and pure activity, as that which exists in virtue of its self-positing and vice versa, as simultaneously agent and product, action and cause of action.[20] The language is not particularly psychological, but it does focus on act, action, and agent. Schelling, however, takes pains not to speak of the I as a subject: if it is called "I," it is at the conclusion of a process of reasoning similar to the "negative theology" of the medievals. The absolute and unconditioned cannot be an object, argues Schelling, for an object is both a thing and something conditioned: bethinged, or limited by other things. But by the same reasoning the unconditioned cannot be a subject either, for subjects "have" objects and their subjecthood is conditioned by their epistemic dependence on an object. To speak of an "unconditioned" subject would be almost as oxymoronic as to speak of an absolute object. Philosophers ought to speak carefully, scolds Schelling, and not fall into blather about the "existence" of God or a "thing greater than which cannot be conceived." To call the unconditioned "the absolute I" may be permitted as a concession to inexact habits of speech, but its sole meaning is *the utterly nonobjective*.[21] When Schelling includes this essay in his *Works* in 1809, after the public break with Fichte, he underscores the "purity" of its conception of transcendental idealism and its lack of contamination by the subjectivism (Fichte's, of course) which later befell philosophy.[22]

Schelling could not (or choose not to) follow the theoretical deductions of the *Grundlage*, for he is not interested at this point in Fichte's precise problem: the objectivity and necessity conveyed by presentation, even when explained from the I's activity and self-positing. He instead chooses to do what Fichte does not, or to do extensively what Fichte does briefly, to characterize the absolute I in terms of categories. If one takes "categories" in the strict sense Kant gave to the term, neither philosopher "categorizes" the unconditioned I. Fichte connects *I am!* with *I think!* or self-positing to explain the self-realization and self-assertion involved in the I's positing, but *being* or *existence* is not a Kantian category. Schelling characterizes the unconditioned I at length, but not in terms of finite categories, e.g., multiplicity, or finite substance, or causality, for these can be applied only to objects or finite things. He does employ the metacategories used to group the twelve: *quantity*, *quality*, *relation*, and *modality*. In general, his approach is negative-theological here, too, as in the basic characterization of the unconditioned principle. He begins to work through the Kantian table in a straight line, e.g., denying empirical unity, plurality, or multiplicity and so concluding to supernumerical unity. His argument soon veers back to Spinoza's *Ethics*, however, and under the metacategory of "quality" it asserts the infinity, indivisibility, and immutability of the I. Under relation, Schelling again follows Spinoza rather than Kant and ascribes to the I absolute immanent causality rather than moral or purposive causality. He treats modality not as a metacategory, but even in considering the triad of possibility, actuality and necessity, he most plainly departs from Kant's guidance. These concepts which Kant thought not real categories, i.e., not strictly objective features of phenomena, but points of view dependent on our perception and judgment of things, Schelling calls the "syllepsis of all categories," or "the sylleptical concepts of all synthesis." A certain Fichtean unification of Kant's table is achieved here, it should be noted, for possibility, actuality, and necessity are interpreted as thesis, antithesis, and synthesis.

If one reviews *On the I*'s argument as a whole, one finds two distinct (not easily reconciled) styles of thought in play. On the one hand, there is the transcendental idealism of Kant and Fichte that demands that the thinkable be limited by the conditions of phenomenal subject-objectivity, and on the other, the metaphysical monism of Spinoza that does not hesitate to characterize the whole as such, or even to take the *principle* for the explanation of entities *inside* experience as itself an item of philosophical investigation. Had they carefully read each other's writings and gotten clear about their own assumptions, Fichte and Schelling would have started bickering much earlier than they do in their correspondence of 1800–1802. Schelling's gradually growing into his "own" system, the system of identity, is largely a matter of him getting clear about his Spinozism, namely, recognizing the logical impossibility of being Kantian *and* Spinozist. Though I have suggested in print that the objective idealisms of Schelling and Hegel were merely an extension of transcendental idealism—on the formal side toward heuristic unification, on the material side, toward a broadened notion of "experience" that included community, social interaction, even world history—I now see such that was *not* the case, however much I wish it were. The identity philosophers snuck around the transcendental in order to return to the transcendent; they exploited the ambiguity of theological language applied to cultural entities to do so, leaving it unclear whether they talked of the absolute whether they were talking of the "One and All" or of a "whole of parts." Our histories of philosophy in their desire to see the *Weltgeist* working in a tidy, linear pattern generally omit the uncomfortable fact that both Fichte and Schelling eventually return to philosophical theism.

III

How much was Schelling prevented from appreciating the Theory of Science by the *Grundlage*'s truncated publication? At the time he wrote *On the I*, he had not seen the section on the foundation of practical philosophy section that was issued at between July and August 1795. Though this lacks a lucid and popular commentary that would coordinate it with earlier sections of the work, grappling with it is crucial for any reader who would understand Fichte's struggle over the congruence of the empirical I, dependent on the not-I in its presentational mode, with the absolute I, stipulated to be self-realizing and active without external limitation. Without a glimpse into the double exorcism of the not-I from the system, once in the deduction of presentation where it is explained as the I's own activity, alienated and hence pictured by the imagination as alien, and again in the deduction of *drive* where difference is seen as primitively inhabiting the self because the self is self-affected or acts against itself as noncausing causality, it is impossible to see that the *Grundlage's* train of thought comes to completion. If Fichte's reader does not follow the hints in section three of "The Concept of Theory of Science" and think along with him how the practical *Wissenschaftslehre* is really the foundation of theoretical, she is likely to misread the "foundations of practical theory of science" as the whole practical philosophy sketched out in the Aenesidemus review, and to think of the *Grundlage's striving* as moral endeavor, the collective historical drift of the human community to realize freedom. Schelling indicates in *On the I* that this is his general understanding of Fichte's philosophy as a whole.[23]

The only thing, however, that is deduced in the *Grundlage's* concluding section is bare will, Leibnizean appetite, the impetus toward the minimal alteration of empirical reality. It is this appetite, or *drive,* that interacts with presentation in that it is drive *to change presentation,* which explains, if anything does, how the empirical I is linked to an objective reality by which it affects itself. In the context of the whole Theory of Science, presentation reduces to will, epistemology to philosophy of action; it is this all-embracing stance of *action within the constraints of empirical finitude and intersubjective limitation* that provides the platform for ethics, and social and legal philosophy.

Schelling is not far from this view of practical philosophy as a whole: his consistent Spinozism drives him to embrace an empirical determinism at the phenomenal level, and to deny the possibility of anything being other than just as it is on the absolute level, where freedom is absolute but no alternatives are possible. The reason he adopts this position, however, is the immanent causality of the unconditional in dependent and conditioned being; he is not yet able to conceive, as he will in the *System of Transcendental Idealism,* that the final locator of the phenomenal individual is the interaction of wills in community, or the self-affection of will as a mutually constraining community of agents.

Schelling's faithfulness to the Fichtean transcendental construction decays at the point in *On the I* where the ultimate Spinozistic metaphysical category of *infinite power* is deployed to explain the causality of the unconditioned upon the conditioned.[24] Had Schelling read the *Grundlage's* conclusion, perhaps he would have been turned away from the arid metaphysical monism of his essay's second half and explored in earnest the affective context into which Spinoza used the idea of power as *conatus* or "endeavor to exist." Schelling in fact shows no great interest in human psychology or morals or philosophy of action, until he abandons the naturalism of his early systems for the spiritualism of his 1809 essay on human freedom.

IV

There are obvious limitations to a comparative study of two philosophers who shared similar visions of the task of philosophy, who work independently but along roughly parallel lines, who read each other's work casually but not fully or in depth. One arrives at no clear linear picture of "causal influence," as if Schelling had wanted to be the devote disciple Fichte took him for, nor at any agonal picture of flatly incompatible positions. This disappoints our dramatic or literary expectations, for a tale ought to be more significant when edited and retold, and a literary dialogue ought to have clear positions and figures, e.g., a Hylas and a Philonous.

Perhaps a historical comparison can bring the work of Fichte and Schelling in 1795 into closer focus. The author of *On the I* and that of the *Foundations* stand to each other as do Spinoza and Leibniz.[25] Spinoza and Schelling share a taste for the metaphysical big picture, and prefer to see substance infinite and will or action finite. Fichte and Leibniz share a taste for the phenomenal, for explanation from the point of view of the perceiver and agent; they share a vitalism as well. Nonetheless Fichte and Schelling (sometimes, for the latter) are post-Kantians and work with the hypothetical-heuristic territory of transcendental supposition, while the pre-Kantian figures acknowledge no in principle intellectual constraints upon their thinking, once the enigmatic Cartesian criteria of clarity, distinctness, and adequacy have been met.

Whether the above comparison is illuminating, I am not sure. If one can recognize, however, that Schelling's construal of transcendental idealism in 1794–95 is metaphysical or Spinozist, that it reifies and distorts the transcendental point of view, perhaps this can shed light on what Fichte was really doing in the *Grundlage*. It is clear that this work does not cash in the broad systematic promises of "Aenesidemus" and the "Concept of Theory of Science," part 3. I have suggested instead that it brings forward a two-part yoked analysis of the structures of action-and-reaction on which any detailed account of human cognition would have to be built: the logical foundation for phenomenologies of perception and of volition.[26] The first part of this analysis, section 4, is a *statics* of finitude, an account of the epistemic dependence of subjecthood on objectivity. It seems to be an idealistic counterpart of the account Spinoza offered of mind as *idea* or reflection of a state of body (or self), or rather of *change of state* in the body (or self). The second part, section 5, is a *dynamics* of finitude. It seems to be a reflection of Leibniz's monad or perception substance that is driven by *appetite*, i.e., by anticipation of *change of state*. The finite subject or empirical I perceives only its own states, or change of states, and its awareness is either coupled with or fueled by movement toward a change of state. Because the I never *is* a state, but is always and only the *process* of changing states, the space between subject and object first opens up and the difference between *having* states and the states that arise and pass away comes to prominence. That opening up of the epistemic and logical-predicative gap is *consciousness*. If this is what Fichte did, we indeed have a deduction of the Reinholdean "facts of consciousness."[27]

NOTES

1. Fichte got a hint that philosophy might be built upon the I from the preacher Johann Schulz of Königsberg in 1791. By 1793–94 he was privately announcing his conviction that the I was both self-realizing and self-thinking. See Manfred Zahn, "Editorischer Bericht" to J. G. Fichte, *Grundlage der gesammten Wissenschaftslehre* (1794–95), in *J. G. Fichte-Gesamtausgabe der Bayerischen Akademie der Wissenschaften*, eds. Reinhard Lauth, Hans Gliwitzky, and Erich Fuchs (Stuttgart–Bad Cannstatt: Frommann-Holzboog, 1964ff), I/2: 176, 177n.

2. Fichte wrote *Ueber den Begriff der Wissenschaftslehre* between February and April 1794, while delivering the outlines of what would become the *Grundlage* in a lecture series at Zurich. Though there was no firm outline of the practical philosophy at this time, the main framework of the theoretical philosophy was in place. Ibid. pp. 179–81.

3. *Grundlage der gesammten Wissenschaftslehre*'s first number, comprising the ground principles and the theoretical philosophy, was published in September, 1794. The rest of the work, chiefly the foundations of practical philosophy, did not appear until the end of July the following year. Ibid., p. 175.

4. See Karl Leondard Reinhold, *Versuch einer neuen Theorie des menschlichen*

Vorstellungsvermögens (1789) (Darmstadt: Wissenschaftliche Buchgesellschaft, 1963), pp. 216–97, and "Ueber die Möglichkeit der Philosophie als strenge Wissenschaft," in *Beyträge zur Berichtung bisheriger Mißverständnisse der Philosophen: Erster Band, das Fundament der Elemetarphilosophie betreffend* (Hamburg: Meiner, 1978), p. 165.

5. See Fichte's comment in the programmatic part 3 of the essay "On the Concept of the Wissenschaftslehre": ". . . It is only in the second part that the Theoretical Part is precisely delimited and given a sound foundation" (Daniel Breazeale, ed. and trans., *Fichte: Early Philosophical Writings* [Ithaca, N.Y.: Cornell University Press, 1988], p. 135). Schelling shows some awareness of this, perhaps, when he writes: "Your empirical I would never strive to maintain its identity if the absolute [I] were not originally posited through itself as pure identity by its absolute power." (*Vom Ich als Princip der Philosophie, Schelling Werke: Akademie Ausgabe* I/2 [Stuttgart: Frommann, 1980], 105.)

6. See *Monadology*, §§ 14–15, 19, 64, 79.

7. ". . . The concept of a causality which is not a causality is, however, the concept of *striving*. Such causality is conceivable only under the condition of a completed approximation to infinity. . . . This concept of striving (the necessity of which has to be demonstrated) provides the foundation of the second part of the Wissenschaftslehre, which is called the Practical Part." "On the Concept of the Wissenschaftslehre," in Breazeale, *Fichte*, pp. 134–35. See also "Review of Aenesidemus," in ibid., pp. 74–76.

8. See Fichte's comments on the method of abstraction and reflection in "On the Concept of the Wissenschaftslehre," pp. 126–27, 132–33.

9. *Faust* I: 339–342.

10. In a letter to Hegel of January 6, 1795, Schelling speaks of contemporary philosophy as oppressed by the dead letter of Kant's text. He quotes with approval Fichte's quip that it requires the genius of a Socrates to figure Kant out, and points to him as the "new hero" on the philosophical scene. The event that occasions Schelling's enthusiasm is his receipt of the first section of *Grundlage der gesammten Wissenschaftslehre*. See Hartmut Buchner, "Editorischer Bericht," *Vom Ich*, pp. 18–20.

11. On July 2, 1795, Fichte writes to Reinhold about the publication of *Vom Ich*. He sees it only as a commentary on his thought; though he is happy it can serve as a vehicle for his being understood by those who cannot understand *him*, he wishes that Schelling would acknowledge its unoriginal origin. Fichte nonetheless pronounces himself pleased with the work, *especially* with its references to Spinoza, whose system is most apt to explain his own. Ibid., pp. 37–38.

12. In a letter to Hegel on February 4, 1795, Schelling replies to his friend's question whether he thinks Kant's moral "proof" for God's existence leads to a personal God. He says he has traded theism for a Fichtean, purely moral concept of deity: "In this respect, I have become a Spinozist. Do not be surprised." He clarifies the remark by explaining that both Kant and Spinoza pose concepts of an absolute, Kant one of the I or its freedom, Spinoza one of an absolute object or

not-I. Ibid., p. 23. Buchner cautions that this text ought to make an interpreter wary of seeing too much Fichte in *Vom Ich*. But he also notes that the Spinoza that Schelling incorporates into this essay is a Spinoza viewed through transcendental lenses (ibid., p. 27).

13. In the letter to Hegel of January 6, 1795, where he praises Fichte as the present-day hero of philosophy, Schelling closes by voicing his determination to provide a modern (transcendental?) counterpart of Spinoza's *Ethics* (Ibid., p. 19.)

14. *Philosophische Briefe über Dogmatismus und Kriticismus* (1795) in *Sämtliche Werke*, hrsg. K. F. A. Schelling (Stuttgart & Augsburg: Cotta, 1856 ff.), vol. 1, pp. 302–305.

15. Ibid., pp. 310–13.

16. In a letter to Reinhold of April 1795, Fichte confesses that the theoretical philosophy is haunted by an intrinsic darkness, which he hopes the practical philosophy will be able to dispel (Zahn, "Editorischer Bericht," p. 185). On July 2, 1795, Fichte writes to Reinhold: "What I want to say is something that cannot be said, nor conceived, but only *intuited*. What I say can do no more than lead the reader to form the desired intuition in him. I would warn him who would study my writings to let words be words, and to seek only to tap into the series of my intuitions, even to keep reading when he does not understand until in the end a spark of light is struck." (Cited in ibid., pp. 216–17). Fichte repeats the same warning to Reinhold when he sends him the practical philosophy in August 1795, saying that the sense of the whole of his philosophy is not to be built up from its individual parts, but rather the reverse: the individual part must be illuminated in and through a sense of the whole (ibid., p. 219).

In correspondence with Goethe, Fichte faults himself for his inability to achieve the lucidity of "intellectual feeling" (ibid., p. 186). By 1801, Fichte finds the exposition of the *Grundlage* darker than it needs be, and complains that the letter, fit to name the thing, kills the spirit (ibid., p. 187).

17. See C. L. Reinhold, "Ueber den gegenwärtigen Zustand der Metaphysik und der transcendentalen Philosophie überhaupt," in *Auswahl vermischter Schriften, Zweiter Theil* (Jena: Johann Maukee, 1797), pp. 331–34.

18. See *Ueber die Möglichkeit einer Form der Philosophie überhaupt*, in *Sämtliche Werke*, vol. 1, pp. 94–96.

19. See ibid., pp. 104–109. In *Vom Ich* § 10, Schelling argues that the categories originate as forms of synthesis between the I and not-I (112n–113n). This section demonstrates a more detailed acquaintance with Fichte's three fundamental principles than does any other passage in the essay.

20. *Johann Gottlieb Fichtes Sämmtliche Werke*, ed. I. H. Fichte (Berlin: Viet & Co., 1845–46), I: 96. Reprinted, along with *Johann Gottlieb Fichtes nachgelassene Werke* (Bonn: Adolphus-Marcus, 1834–35), as *Fichtes Werke* (Berlin: de Gruyter, 1971).

21. *Vom Ich*, §. 3, AA I, 2: 89–90.

22. Ibid., p. 81.

23. In an enthusiastic letter to Hegel on February 4, 1795, Schelling pictures the relation of theoretical philosophy and practical philosophy this way: in theoretical philosophy an infinite sphere is divided up into many finite spheres by the positing of limits. A contradiction ensues between the finite and the infinite, and it is suspended only with the breakthrough into the infinite that the practical effects. The practical stance

> demands the destruction of finitude and so transports us to the supersensible world. (Practical reason does what theoretical reason cannot, since it is enfeebled by objects). But we find in the supersensible nothing other than our absolute I, for it alone describes the infinite sphere. There is no supersensible world for us other than the absolute I.

Cited in Buchner, "Editorischer Bericht," pp. 22–24.

24. Section 14 makes clear that ascribing absolute power to the I abolishes the supposed ability of a finite mind to act for the best. The rule of wisdom is suspended in favor of the determinations of force (*Vom Ich, Akademie Ausgabe*, pp. 122–23). To this he joins a spirited polemic against his theological instructors, the "seminary" Kantians. Kant's notorious postulates of God's existence, willingness to reward merit with happiness, and of the endless duration of soul have nothing to do with morality, which is simply the unconditioned command that the limited I become the absolute I. Were this in fact possible, the moral law would be suspended as *obligatory* and instead become a *law of nature* (ibid., pp. 125–26).

25. Schelling makes clear his admiration of Spinoza and his wish to combine certain features of Kantianism and Spinozism in the close of the preface to *Vom Ich*. Though it is at least programmatically clear that for the Kantian philosophy the whole essence of the human is freedom, Schelling thinks that to date this had been worked out only in fragments. He voices the hope that he can produce a *counterpart* to Spinoza's *Ethics* along this line. *Vom Ich, Akademie Ausgabe*, I, 2: 78, 80. Schelling later attempts to formulate a "system of freedom" in the 1809 essay on human freedom.

26. That there is a logical model of action and reaction for all psychic events is a Leibnizean insight: an immaterial substance is defined as one that contains force and perception, "force" being the principle of change or action. "On the Supersensible Element in Knowledge, and on the Immaterial in Nature," in *Leinbniz Selections*, ed. P. Weiner (New York: Schribners, 1951), p. 354.

27. The same opening of a gap explains, i.e., provides a necessary but not a sufficient condition for, the *reflection* Spinoza posited between *idea* and *ideatum*; this reflection (based on the registering of changes of state) itself explains self-awareness: the fact that when one has an idea one can also have an idea of the idea. See *Ethics* 2: P13Dem, P16Cor2, P19Dem, P21S.

BETWEEN KANT AND HEGEL
Fichte's *Foundations of the Entire Science of Knowledge*

Vladimir Zeman

Fichte's own philosophy has recently become a subject of more concentrated attention than ever before. But even when we no longer consider him primarily as the main mediating link between Kant and Hegel, this problem area is still worth further consideration. The purpose of this paper is to deal with a selected aspect of this problematic. Of central interest for us will be the following three claims made by Fichte in his first Jena *Wissenschaftslehre*.[1]

(1) That what he presents is critical philosophy:

> Now the essence of the *critical* philosophy consists in this, that an absolute self is postulated as wholly unconditioned and incapable of determination by any higher thing; and if this philosophy is derived in due order from the above principle, it becomes a Science of Knowledge.[2]

(2) That his philosophy is transcendental idealism:

> Transcendental idealism thus appears at the same time as the only dutiful mode of thought in philosophy.[3]

In the "Second Introduction" Fichte characterizes transcendental idealism in general as being based on an assumption that[4] all consciousness rests on, and is conditioned by, self-consciousness.

(3) That his philosophy as critical philosophy and transcendental idealism has that systematic form that lends itself better to the presentation of Kant's philosophical project than the one used by Kant himself:

In the critical system, a thing is what is posited in the self . . . critical philosophy is thus *immanent*, since it posits everything in the self . . . . According to the Science of Knowledge, all consciousness is determined by self-consciousness. . . . Now I am very well aware that *Kant* by no means *established* a system of the aforementioned kind . . . . However, I think I also know with equal certainty that *Kant envisaged* such a system. . . .[5]

In what follows, these three claims will be presented and critically analyzed, primarily through comparison with similar claims by Kant, whose project Fichte claimed to realize properly, as well as with Hegel's views from the time of the *Differenzschrift*.

The title of this paper might look rather presumptuous unless some qualifications are added.[6] First, Fichte's positive project of a new form of transcendental philosophy is not on all counts a direct critical response to Kant. If we took a primarily historical approach, at least Reinhold, Schulze, Beck, and Maimon would have to be considered as well. Frederick C. Beiser characterizes the situation faced by these and other early Kantians as follows:

If one were a Kantian in the early 1790s, the main question was no longer how to defend Kant against his enemies, but how to rebuild the critical philosophy from within upon a new foundation. . . . Reinhold's meta-critical methodology—his ideas concerning the proper method of transcendental philosophy—gained wide influence and became virtually canonical for the post-Kantian generation. Three ideas of his in particular were assimilated by Fichte, Schelling, and Hegel: (1) the demand that philosophy be systematic; (2) the insistence that philosophy begin with a single, self-evident first principle; and (3) the claim that only a phenomenology can realize the ideal of a *philosophia prima*.[7]

The importance of such "mediation" between Kant and other main representatives of German classical philosophy was recognized relatively early. In 1840, an excellent historian of philosophy of Hegelian orientation (as well as coeditor of Kant's collected works) Karl Rosenkranz, wrote:

Fichte accomplished what Maimon, Reinhold and Beck tried, namely to deduce from one point all individual principles which in Kant still remained unconnected, as well as not to have categories simply to arise.[8]

It is debatable to what degree it was actually Reinhold's *Satz des Bewusstseins*, proposed in 1790[9] that already accomplished such task. However, if nothing else, Reinhold at least spelled out the characteristics to be met by any such principle:

(1) It must be the foundation, the sufficient and necessary reason, for all other true propositions. . . .

(2) Its terms must be precise and self-explanatory. . . .

(3) The first principle must be of the highest generality, so that its terms are the most universal concepts of which all others are only species. . . .

(4) [and most importantly] The first principle must also be a self-evident or immediate truth. More precisely, it cannot require any reasoning to be found true; for, as the first principle of all demonstration, it cannot itself be demonstrated. The proof of the first principle must therefore lie outside the science that it is to demonstrate.[10]

Unfortunately, the complexity of the real historical development can only be outlined here. Rather than dealing with various forms of transcendental philosophy, I prefer to limit this discussion to Beiser's first claim, i.e., the difference in character of what all these three men call *system* of transcendental philosophy in general, and how they try to justify it in particular.

Each of the three philosophers central to our project not only deals in his own particular way with the "scandal of metaphysics," but also claims to have overcome it by establishing a new system of scientific philosophy. In one way or another, in their time all three claimed, and have since been seen by others, to "close" the evolutionary line of philosophy. However, historically, all three at the same time opened in their respective works new ways for doing philosophy that went beyond the conceptions they in effect authorized. Already within the scope of what we learned to call German Classical Philosophy, the term "transcendental philosophy" came to represent more a certain way of doing philosophy rather than simply a name for one particular system. As an ironic consequence of its popularity, this term then became devalued in the same fashion as has happened to other historical characterizations of new ways of doing philosophy.[11]

It was the historical destiny of the Kantian philosophy that, with the intention and belief to limit metaphysics critically, it in truth lent a new strength and a new resonance to the ultimate spiritual grounding motives of metaphysics. For the "Kritik der reinen Vernunft" not only delivered empirical but also metaphysial cognition from the restraints of the dogmatic concept of a thing. Now metaphysics, in order to fullfil its tasks, no longer had to be knowledge of absolute things which exist transcendentally "outside" the mind; its true goal lay rather in the complete conception of the organization of the mind itself. The great speculative system formations differ henceforth according to the assumptions and the starting point from which they try to determine this organization.

I intend to focus primarily on Kant and Fichte and to propose that, metaphorically speaking, the centers of gravity, forms of justification, and logical forms of their respective systems vary to such a degree that, while they themselves claimed to have been operating within the sphere of transcendental philosophy, (a) it was not without justification that they did not see each other's philosophy in such a way, and (b) their philosophies were not necessarily viewed as transcendental by other philosophers.

In Kant[12] we find *both* metaphysical and transcendental argumentation. For the legitimation of philosophy is not reduced to its a priori character, or its derivation from some "self-standing" theoretical principle. These two directions of transcendental argumentation run respectively from the transcendental framework of all possible knowledge to actual knowledge and from actual knowledge to the transcendental framework (B737). It might be useful to remind ourselves of the very beginning of the Introduction to the *Critique of Pure Reason*:

> There can be no doubt that all our knowledge begins with experience.
> . . . In the order of time, therefore, we have no knowledge antecedent to experience, and with experience all our knowledge begins. But though all our knowledge begins with experience, it does not follow that it all arises out of experience. (B1)

Since Kant considers everyday scientific knowledge as well as philosophical knowledge in a certain sense as a part of the same cognitive continuum, he refuses to consider any nonlimited (and in a material sense absolute) knowledge, be it divine or some other. In the case of nonformal knowledge, he works after the 1770 dissertation solely with the concept of sensible but not intellectual intuition. Kant's division of reason into theoretical and practical treats the "I" in the cognitive framework as a purely logical principle, which plays no generative role.

However, the most interesting point, at least in our context, is the type of the system he views as appropriate for philosophy. Since Kant attempts to "define" or delineate a priori the *framework* of possible experience as well as to show that experience actually fits such a framework, he cannot follow too closely the example of mathematics, or to be more precise, of geometry—in his view, mathematics constructs its concepts while philosophy has to handle them as it encounters them.

According to Kant,

Philosophical knowledge is the *knowledge gained by reason from concepts*; mathematical knowledge is the knowledge gained by reason from the *construction* of concepts. To *construct* a concept means to exhibit *a priori* the intuition which corresponds to the concept. For the construction of a concept we therefore need a *non-empirical* intuition. (B741)

Yet also according to Kant,

Whereas . . . mathematical definitions *make* their concepts, in philosophical definitions concepts are only *explained*. . . . In short, the definition in all its precision and clarity ought, in philosophy, to come rather at the end than at the beginning of our inquiries.(B758-59)

Philosophy cannot strive toward axiomatization in a strict sense. Principles of pure understanding are neither of an empirical nor of a formally logical character and origin; metaphysical deduction is necessary but not sufficient for Kant's principles. In the same way, ideas in their properly regulative use are concepts with a maximal "ordering" power exercised over the "lower" level concepts of understanding. In accordance with the famous sequence of questions in the *Prolegomena*, metaphysics still belongs to the same continuum of sciences, based on synthetic a priori judgments. Yet unlike mathematics, it does not proceed through derivations. Metaphysics shares the framework for its operation with what Kant calls the pure science of nature. However, metaphysics differs from pure science of nature in that it cannot strive for even a partial mathematization of its procedures and results. Most importantly, the bases for both the pure science of nature and scientific metaphysics present themselves in a form of interrelated principles, rather than as one principle only.

This twofold, or bidirectional, task given in Kant's conception of the framework of possible experience, serves Kant as a criterion of demarcation between purely speculative and scientific metaphysics. While Kant's discussion of the impossibility of an ontological proof of God's existence is usually considered within the limits of this special problem only, it provides us with a more "universally" applicable warning as well.

[T]he illusion which is caused by the confusion of a logical with a real predicate (that is, with a predicate which determines a thing) is almost beyond correction. Anything we please can be made serve as a logical predicate; the subject can even be predicated of itself; for logic abstracts from all content. But a *determining* predicate is a predicate which is added to the concept of the subject and enlarges it. Consequently, it must not be already contained in the concept. (B626)

However, what is to be analyzed here is Fichte's and not Kant's philosophy. It would be false to blame Fichte for attempting a *derivation* of transcendental philosophy from some *supreme* theoretical principle, as was the case in traditional philosophy. In his attempt at founding philosophy independently of and prior to experience, he concentrates on originative and universally unifying pure *practical* activity. Let us return once more to our relatively neutral party, Karl Rosenkrantz.

> Philosophy could progress from two points only: from self-consciousness and from the thing in itself. It had to arrive at the unity of subjectivity and objectivity which was in Kant just a relation, assertoric and problematic. *Self-consciousness* had to set itself as producing on its own everything objective and thus positing it against itself as something completely different, but which already includes it as its *own otherness*.[13]

The purely logical character of the "I think" had to be replaced by a conception in which the "I" would be both ideal and real.

> According to Kant, the unity of all acts of intelligence is the synthetic apperception *a priori*, the I think. But Kant conceived this unity only as formal and not as real *as well*. He had put the aesthetic and logical elements together in an external way only. Fichte recognized the unscientific character of such incoherence. According to Kant, the I should accompany all acts of intelligence, but only to accompany them, i.e. to remain external in their respect. However, since thinking is only as being and being is only as thinking, it is thus the negation of the difference between the intuition as the principle of the real knowledge and the understanding as the principle of the ideal knowledge, the intuitive and supraintuitive knowledge. *The I*, the concept of self-consciousness, is the *principle of science*. The opposition of intuiting and thinking set by Kant, comes together for him as *intellectual intuition*. As a *subject*, the I is not without object, but contains within itself its own *object*; it is an I only so far as it is for itself. The I is thus the *ideal and real principle*.[14]

In the case of the I and the proposition expressing the first absolutely unconditioned principle, Fichte indeed appeals to Kant's authority, creating an impression, a false one, I believe—given Kant's own claims about methods of philosophy, that Kant was moving in the direction of such a concept of a philosophical system.

> That our proposition is the absolutely basic principle of all knowledge, and it was pointed out by *Kant*, in his deduction of the categories; but he never laid it down specifically *as* the basic principle.[15]

To return to Fichte's construction of his own philosophy, both pre-Kantian dogmatic and Kantian critical idealism were to be replaced by practical idealism. Fichte writes in his *Second Introduction to the Science of Knowledge*:

> Intellectual intuition is the only firm standpoint for all philosophy. From thence we can explain everything that occurs in consciousness; and moreover, only from thence. Without self-consciousness there is no consciousness whatever; but self-consciousness is possible only in the manner indicated: I am simply active. Beyond that I can be driven no further; here my philosophy becomes wholly independent of anything arbitrary, and a product of iron necessity, insofar as the free reason is subject to the latter: a product, that is, of *practical* necessity. I *can* go no further from this standpoint, because I *may* not go any further; and transcendental idealism thus appears at the same time as the only dutiful mode of thought in philosophy, as that mode wherein speculations and the moral law are most ultimately united. I *ought* in my thinking to set out from the pure self, and to think of the latter as absolutely self-active; not as determined by things, but as determining them.[16]

The concept of action, which becomes possible only through this intellectual intuition of the self-active self, is the only concept that unites the two worlds that exist for us, the sensible and the intelligible.

On the issue of intellectual intuition, even Ernst Cassirer, a leading neo-Kantian, who could scarcely be suspected of neutrality, adhered to the view that since it is not located in the cognitive sphere, it does not lead to a break from critical philosophy. "However, the 'intellectual intuition' is according to Fichte's own basic view never an intuition of something what is but of some activity."[17] Unfortunately for Fichte, not everybody shared this positive view; if intellectual intuition is both outwardly and inwardly oriented, then at least the *Sein* of the I, or I in its being is within its scope. As a consequence, Fichte can be thus seen as developing at best another a new form of transcendental philosophy. Yet it may still be disputed whether either what he aimed at and what he achieved was still a form of critical philosophy, as Fichte himself claims at the end of the preface to *Foundations*.[18] Once more Rosenkrantz,

> [I]t depends on *balancing* what is determined, what exists, what is not conscious of itself, with the I. Therefore, Fichte did away with the various starting points of the Kantian *Critiques*, expressed in the simplified opposition which determined, on the *theoretical* side, the I by the not-I, subject by object, thinking by being. But on the *practical* side the not-I was determined the other way around by the I, the object by the subject, being by thinking. [And most importantly:] In this way, transcendental idealism incorporated the otherness of the thing in itself into the *a priori* synthetic unity of self-consciousness, and even though it might not yet been overcome, it already became powerless.[19]

Methodologically, Fichte seems to have adhered to a view still common in his time, that for the sake of the highest possible certainty, philosophy has to be capable of deriving its whole system from the absolutely self-supporting Archimedean point, a "proposition that everyone will grant us without dispute."[20] Anything less would be a case of *Halbheit*! In this sense, Kant's philosophy had to appear as unsatisfactory. Accordingly, only such science of knowledge, which was based on "the primordial, absolutely unconditioned first principle of all human knowledge"[21] could make a claim to determine absolutely all knowledge and science both in its form (therefore its logic as well) and content. While Kant and Fichte could both be criticized for their ahistorical view of science, in Fichte's case even the "feedback" side of Kant's transcendental philosophy (feedbacks and checks) was lost in the process of the further development of transcendental philosophy, and the corresponding structure of the system was abandoned. From this particular angle, it would appear that we have simply another specimen of a purely metaphysical system, based on one principle only, to which no system existing in positive science did, or it seems, could correspond. And at least in this latter sense, it would be hard not to view it as another (at least partly unplanned) return to a dogmatic, pre-Newtonian and pre-Kantian view of systematic structure.

In the case of Hegel's philosophical theory, as viewed in his early work, one should ask whether and in which sense it is of transcendental character, in particular when compared with Kant's and Fichte's. First, the framework of possible experience enabling the determination of objects of empirical experience has been here replaced. According to Cassirer,

To be sure, even for *critical idealism* understanding is the "creator of nature";—it, too, explains the possibility of a priori through the claim "that we know about the things only that which we have ourselves put into them." However, the spontaneity of understanding does not signify here the spontaneity of production but of determination.[22]

In respect of this *"Erschaffen"* (normally translated as "creation" or "production"), Hegel's and Fichte's philosophy both represent a species of positive and material a priori, oriented toward metaphysics rather than toward critical or formal idealism.

Second, it must be asked in which sense Hegel's philosophy can be considered a form of transcendental philosophy. According to the *Differenzschrift*,

> The method of the system, which can be called neither synthetic nor analytic, occurs in its purest form when it appears as a development of Reason itself. . . . A genuine speculation . . . proceeds necessarily from absolute identity. Its diremption into subjective and objective aspects is a production of the Absolute. The ground principle is thus completely transcendental. . . .[23]

And third, what is the logical structure of Hegel's system? Franz Kroener called it a system of systems and subjected it to considerable criticism from the standpoint of formal logic.[24] In our post-Russellian period, we could possibly consider it as a fallacious attempt to create an analogue of the catalogue of all catalogues.

As to its form and the need to identify a starting point, Hegel's system is closer to Fichte's than to Kant's. What it provides is at the same time a foundation, as well as a transcendental framework, for all aspects of human existence. If, in the words of F. H. Bradley, the system has become for Hegel "an arbiter of fact," then systemacity has been "transmuted" from a *"hallmark of science* into *a standard of truth"* and Rescher is right to speak about a "Hegelian Inversion."

> If a thesis coheres systematically with the rest of what is known, then— and only then—it is a part of real knowledge (which accordingly characterizes reality itself).[25]

Except for the short phrase in brackets in this passage, what Rescher refers to as an "Inversion" in principle allows for a nonontological interpretation. However this was not Hegel's own position. Nevertheless, and here is an important difference from Fichte, its derivative structure is not linear. More metaphorically, its general shape corresponds to a spiral, deterministic model with clearly defined starting and final points, which is further enriched by the dialectic type of conceptual cum objective progress. In that the strategy for conceptual justification is not linear but circular, such circularity is different from Kant's. For it is strictly contained within philosophy itself and without consideration of "the facts" of either cognitive, moral or aesthetic "experience."

CONCLUSION

In 1929 Franz Kroener published a book entitled *Die Anarchie der philosophischen Systeme*.[26] Some twenty years ago Stephen Koerner published his *Categorial Frameworks*.[27] These two works, as well as those of various other authors (e.g., Nicholas Rescher's work mentioned above) deal with what can be loosely called the "logic of philosophy." In more precise terms, we could speak here about the study of the conceptual logic of philosophical systems.

Without analyzing systematically the results of such studies, I would like to stress at least four relevant points:

(1) It is profitable to identify the type of systematic structure used by a given philosopher (at the risk on occasion of needing to specify in which of his works). Such analysis provides us with a chance not only to differentiate between logical and nonlogical moves initiated by the author, but also to clarify the status of such further issues as conceptual closure, as well as tacit assumptions and their consequences. In respect of the logic used, Seebohm, in his 1976 article on "The Grammar of Hegel's Dialectic," concluded,

> What must be said from the viewpoint of modern logic against Hegel's dialectic is: the doctrine of concepts is not the only possible system, it is even a poor system. Hence assumptions of the type of Hegel's speculative assumption can be made with regard to other systems as well. But this would mean: Hegel's dialectic is not the only possible dialectic and there is a possible manifold of possible explications of the absolute. It can be added, furthermore, that on pure syntactical grounds, the constructivist and *Kantian* viewpoint has a good chance of withstanding the proposals of the monstrator of *Phenomenology*. . . . It should be clear, on the other hand, that any attempt to develop out of Hegel's dialectic a "mate-

rialistic" or "real" dialectic which deals with reality and is not a dialectic of thought form has logical implications just regarding thought form. The first consequence would be to say farewell to the doctrine of concepts. This implies, however, that any reference to Hegel and his dialectic, which might be used in order to explicate these new dialectics, has at best the value of a metaphor.[28]

(2) Kroener already realized that while in some aspects the historically later systems might be derivative (he speaks about the ways in which early Fichte and Hegel presuppose Kant and can be seen as derivative),[29] in some other aspects they are opposed to each other, e.g., in their exploitation of alternative strategies for the construction of a system as well as material (i.e., nonformal) foci. In this sense, any statement about one philosophy *overcoming* another one should be always made more precise by identifying the areas within the system where such comparison is at all possible. In this respect, it would be useful to develop a looser version of the mathematical concept of an invariant.

Historically, the turn of various originally right Hegelians (Fischer, Zeller, Weiss) from Hegel to Kant from the mid 1850s on, can be sufficiently explained by their realization of the structural and explanatory limits of Hegel's philosophical system as well.

(3) If a given philosopher claims to provide a framework or a firm foundation for knowledge and science in general, we should not only take him at his word but verify whether the degree he really succeeded in demonstrating his claim. Kant, who tried to do so in his *Metaphysical Foundations of Natural Science*, was only partly successful; in spite of various simplifications and hasty generalizations, Hegel's *Philosophy of History* provided a great service not only to the philosophy of history but also to history. As to Hegel's excursions into natural science, while philosophers may still discuss their value, such excursions have always appeared to be of little interest for natural scientists.

(4) Every traditional, i.e., universalistically oriented, philosophical system with absolutistic claims already contains in its own logical program a self-destructive contradiction.

[B]ecause it has to select, omit, and determine "what is important", every philosophical system must perish due to the conflict between its claim to cover everything and the limitation of its content. This limitation cannot be eliminated even through the consecutive systematizations of the same basic idea, which would adapt it over and over again.[30]

In the case of Hegel's philosophy, one of those destructive points is the claim that this philosophy has to make about itself. To that degree to which it aspires to be the apex of philosophical development, Hegel's philosophy limits and therefore negates the universal validity of its fundamental element—the dialectic.

I will close with two short final points: First, like everybody else, Fichte does not always do what he claims. However, here I concentrated on his claims, because I consider them crucial for his conception of transcendental philosophy. In the same way, when Kant claims that his system should be able to handle all philosophical problems, this claim may not correspond to what he did or could do. Second, even in the case of Fichte, there seems to be a price to be paid if one aspires not to be just "ein dreiviertel Mann."

NOTES

1. Unless stated otherwise, all references to Fichte's work are based on *Fichte: Science of Knowledge (Wissenschaftslehre) with First and Second Introductions*, ed. and trans. Peter Heath and John Lachs (New York: Appleton-Century-Crofts, 1970).

2. Ibid., I, 119–120, 117.

3. Ibid., I, 467, p. 41.

4. Ibid., I, 462, p. 37.

5. Ibid., I, 120, p. 117; I, 477–78, p. 50.

6. The most recent historical treatment of the "bridge" from Kant to Fichte by Frederick C. Beiser (see *The Fate of Reason: German Philosophy from Kant to Fichte* [Cambridge: Harvard University Press, 1987]) is for our purposes unsatisfactory on two counts: (1) It mentions two "bridgeheads" only in the title; (2) The "from . . . to" is presented primarily in a temporal sense, rather than as a developmental one; it concentrates on philosophical development parallel to the line mentioned in the title.

7. Ibid., p. 227–28.

8. Karl Rosenkranz, *Geschichte der Kant'schen Philosophie* (Berlin: Akademie-Verlag, 1987), p. 375.

9. Primarily in *Beytraege zur Berichtigung der bisherigen Missverständnisse der Philosophen*. Beiser (*The Fate of Reason*, pp. 252–53) translates the first sentence of *Beytraege I* as follows: "In consciousness, the representation is distinguished from, and related to, the subject and object, by the subject."

10. Ibid., pp. 244–45.

11. Ernst Cassirer, *Das Erkenntnisproblem. Dritter Band: Die nachkantischen Systeme* (Darmstadt: Wissenschaftliche Buchgesellschaft, 1974), p. 285.

12. Immanuel Kant, *Critique of Pure Reason*, in N. K. Smith's translation, any edition. Parenthetical citations in the text refer to this work.

13. Rosenkranz, *Geschicte der Kant'schen Philosophie*, p. 371.

14. Ibid., p. 376–77.

15. *Fichte: Science of Knowledge*, I, 99–100, p. 100.

16. Ibid., I, 467–78, p.41.

17. Cassirer, *Das Erkenntnisproblem*, p. 140.

18. *Fichte: Science of Knowledge*, I, 90.

19. Rosenkranz, *Geschicte der Kant'schen Philosophie*, p. 377.

20. *Fichte: Science of Knowledge*, I 92, p. 94.

21. Ibid., I, 91, p. 93.

22. Cassirer, *Das Erkenntnisproblem*, p. 364.

23. G. W. F. Hegel, *The Difference Between the Fichtean and Schellingian Systems of Philosophy*. trans. J. P. Surber (Atascadero, Calif.: Ridgview, 1978), pp. 32–33.

24. Franz Kroener, *Die Anarchie der philosophischen Systeme* (Leipzig: Meiner, 1929), particularly p. 150ff.

25. Nicholas Rescher, *Cognitive Systematization* (Oxford: Blackwell, 1979), pp. 34, 37.

26. See n. 12 above.

27. S. Koerner, *Categorial Frameworks* (Oxford: Basil Blackwell, 1970).

28. *Hegel-Studien*, band 11 (Bonn: Bouvier, 1976), p. 180. I would like to thank Jere Surber for bringing this article to my attention.

29. Kroener, *Die Anarchie der philosophischen Systeme*, pp. 89–90.

30. Ibid., p. 156.

JACOBI'S PHILOSOPHY OF
FAITH AND FICHTE'S
WISSENSCHAFTSLEHRE 1794–95

Curtis Bowman

I t is a commonplace that F. H. Jacobi's writings exercised a great deal of influence on Fichte's thought. In this paper I intend to discuss certain facets of the complex relationship between these two thinkers in an effort to shed light on the origins and reception of the *Foundations of the Entire Wissenschaftslehre*. Most of my paper will be devoted to discussing Fichte's enthusiasm for and appropriation of ideas in Jacobi's writings from the mid-1770s to the early 1790s, just prior to the composition and publication of the *Wissenschaftslehre*.[1] But I shall also briefly discuss both Jacobi's famous letter to Fichte from 1799 and Fichte's rather oblique reply to it in *The Vocation of Man*. So, although this paper is a rather breathless march through many texts, I hope to gain greater insight into the *Wissenschaftslehre* by investigating Jacobi's relationship to this complex work.

During the period when he was composing and lecturing on the *Wissenschaftslehre*, Fichte sent two revealing letters to Jacobi. Together they offer us an excellent source of evidence for the nature and extent of Jacobi's influence on Fichte, since they tell us how Fichte himself perceived that influence. The first letter is dated September 29, 1794, and in it Fichte makes an amazing confession:

> If there is any thinker in Germany with whom I wish and hope to be in agreement in my particular convictions, it is you. . . .[2]

This remark is truly astounding because it was made while Kant was still alive. Given that at this time Fichte was constantly announcing his allegiance to the spirit if not the letter of the Critical Philosophy, it is unsettling to read that he most wanted to be in agreement with Jacobi, who had by this time publically attacked Kant. If it were not for the second letter, we might simply dismiss the first as a transparent attempt to curry favor, for it accompanied the first two parts of the *Wissenschaftslehre*, which Fichte said he was sending to Jacobi as proof of his great esteem for him. But the second letter, dated August 30, 1795, which accompanied the third part of the *Wissenschaftslehre*, indicates that in his first letter Fichte was expressing his genuine regard for Jacobi. Fichte once again expresses his admiration for Jacobi and attempts to explain certain features of the *Wissenschaftslehre* to him at some length:

> This summer, in the tranquility of a charming country house, I reread your writings, and then read them again and again. At every turn—but especially in *Allwill*—I am astonished by the striking similarity of our philosophical convictions. The public will scarcely credit this similarity, nor perhaps will you. But your perspicacity makes me dare to hope that from the uncertain outline of the beginnings of a system you will be able to deduce the entire system. Everyone knows you are a realist and I am, after all, a transcendental idealist and an even stricter one than *Kant*. Kant clings to the view that the manifold of experience is something given—God knows how and why. But I straightforwardly maintain that even this manifold is produced by us through our creative faculty. Allow me to explain this point to you here in this letter.
>
> My *absolute I* is obviously not the *individual*, though this is how offended courtiers and irate philosophers have interpreted me, in order that they may falsely attribute to me the disgraceful theory of practical egoism. Instead, *the individual must be deduced from the absolute I*. . . . As soon as we regard ourselves as individuals—which is how we always regard ourselves in *life*, and only in *philosophy* or in *poetry* do we regard ourselves differently—we find ourselves at that standpoint which I call the *practical* standpoint. (I call the standpoint of the absolute I the *speculative* standpoint.) According to this (practical) point of view, a world exists for us which is independent of us and which we can do no more than modify. From this standpoint, the pure I is posited outside of ourselves and is called God. . . .
>
> What is the purpose of the speculative standpoint, and indeed of philosophy as a whole, if it does not serve life? If mankind had never tasted this forbidden fruit, it could dispense with all philosophy. But mankind has an innate desire to catch a glimpse of that realm which transcends the individual—to view this realm, not merely in a reflected light, but directly. . . .

Allwill allows transcendental idealists to hope for peace and even for a sort of alliance [with realists], so long as idealists are content to protect their own boundaries and to make these quite clear. I believe that I have already met this condition. If I were to go even further, and were I, from within the supposedly hostile territory [of idealism], to guarantee the security of realism's domain, then by rights I should be able to count upon—not merely an alliance of a sort—but upon an alliance in every respect.[3]

This second letter is remarkable for many reasons. Three, among others, stand out at this point in this paper: (1) Fichte's expressed hope that Jacobi would go on to deduce a system of thought from first principles; (2) his claim that there are many similarities between his views and Jacobi's, which seems odd at first glance, considering that Fichte was an idealist and Jacobi a renowned realist; and (3) his emphasis on the importance of *All-will*, a not especially philosophical novel of letters, both for establishing in his mind the similarity of their views and for hoping for the possibility of forging an alliance between the two of them. In what follows I intend to take up these three topics in order to go some way towards clarifying the relationship between these two men and their philosophies.

Philosophers routinely tell one another that they admire and value their works, and they also encourage one another to pursue new lines of thought and develop old ones. Thus Fichte's hope that Jacobi would develop his ideas is hardly surprising, but that he further hoped that Jacobi would deduce a system of thought is perplexing to anyone acquainted with Jacobi's writings. Jacobi routinely exhibits hostility toward systematic thinking, for he believes that it leads us down paths that we should not follow. (This should not be seen as hostility towards argumentative or critical thinking, for Jacobi routinely argues for his views and criticizes others' views.) That is, he sees philosophers's efforts to begin with first principles and then to infer conclusions from them as in the service of a vision of God and man which he violently disapproves of. His suspicion about systematic thinking is best seen in the exchange with Moses Mendelssohn from the early 1780s over the proper interpretation of G. E. Lessing's confession of Spinozism, now known as the pantheism controversy, but we will see that he has similar problems with Fichte's *Wissenschaftslehre*.

The details of the pantheism controversy lie outside of the scope of this paper, but suffice it to say that Jacobi repeatedly argued that the systematic application of the principle of sufficient reason—the supreme example of which was to be found, he maintained, in Spinoza's *Ethics*—led inexorably to atheism and fatalism. Spinozism is atheism, he argued, because the conception of God which it advocates is an immanent one: that is, it does not conceive of God as a transcendent cause of the universe.[4] But an immanent God is simply the material universe. So, for Jacobi, Spinozism is materialism. And Spinozism is fatalism because the universal applicability of the principle of sufficient reason implies that we are not the spontaneous causes of our actions.

But did these conclusions, which Jacobi maintained were sound and valid results of the principle of sufficient reason, cause him to despair? No, for he extricated himself from these difficulties through a *salto mortale*, a fatal leap, in which he rejected the conclusions of reason and accepted those of faith, which told him that there is a God and that we are free. He then summarized his philosophy of faith in a few often-quoted words:

> In my judgement the scholar's greatest merit is to unveil and reveal existence. . . . Explanation is to him a means, a way to the end—the proximate, but never the ultimate goal. His ultimate goal is that which cannot be explained: the insoluble, the immediate, the simple.[5]

There are some things which we cannot explain. We must simply accept them for what they are as we intuit them, and for Jacobi these include the knowledge that God exists and that we are free.[6]

Thus his hostility to systems arises from his belief that they attempt to explain the inexplicable, and that their results contradict those of faith. Hence the oddity of Fichte's hope that Jacobi would go on to develop a system deduced from first principles. But Jacobi's openness to the inexplicable surely appealed to Fichte, and was probably the source of the hope that he would deduce a system from first principles.

I will return to this theme below, but now I would like to turn to the similarities which Fichte claims to find between his works and Jacobi's. There are two which I think are especially helpful in understanding the origins of the *Wissenschaftslehre*: (1) Jacobi's critique of Kant's notion of the thing in itself (found in the appendix of *David Hume* from 1787) and (2) Jacobi's rehabilitation of the relationship between the conditioned and the unconditioned (from the seventh appendix added to *Über die Lehre des Spinoza* in 1789). Both of these influenced Fichte's thinking about the absolute I.

If someone knows anything at all about Jacobi, if only by hearsay, then it is usually that he famously criticized Kant's notion of the thing in itself. Jacobi claims that the thing in itself violates the spirit of the Critical Philosophy, but that nevertheless it is required for Kant's system to be possible. His argument is as follows. First, he observes that empirical objects, which are only appearances, cannot be the cause of our representations, for, according to Kant, they simply are organized collections of representations. There is instead a transcendental object, a thing in itself existing independently of our representations, which is the cause of these representations. But this transcendental object can never be given in experience, for whatever is not an appearance can never be given in experience.[7] This means that Kant is simply assuming the transcendental object to be the intelligible cause of our representations, which corresponds to sensibility as receptivity. That is, sensibility is essentially passive or receptive. Thus there must be something which affects it. The very term *Sinnlichkeit* implies a medium between two real things, between appearances and things in themselves.[8]

But this situation puzzles Jacobi and leads him to make a famous confession:

> I must confess that this difficulty has hindered me not a little in my study of the Kantian philosophy, so that for several years in a row I had to begin the *Critique of Pure Reason* from the start again and again, because I continually began to doubt that without this presupposition I could not enter into the system, and that with it I could not remain in it.[9]

The presupposition in question is, of course, that of the thing in itself. Kant needs to assume it, since sensibility is passive and must be affected by something. But, Jacobi claims, he may not assume it because to do so is to violate the restrictions which transcendental idealism places on our knowledge. The principle of sufficient reason does not apply to the realm of things in themselves, and thus Kant cannot assume that there is a thing in itself which affects sensibility and produces our representations.[10]

Jacobi claims that the transcendental idealist is left with two choices. On the one hand, he may eliminate the thing in itself from the Critical Philosophy, thereby adopting the position of speculative egoism.[11] This is a more consistent idealism in that it admits only our representations into our ontology. Or, on the other hand, he may cease to be an idealist altogether by adopting Jacobi's realism and the philosophy of faith which supports his realism. This second option is not explicitly offered, but it is implicit in what Jacobi is doing. The unpalatability of Kant's idealism, once demonstrated, is supposed to drive everyone into the arms of Jacobi's philosophy of faith.

But Fichte takes the first route, eliminating the thing in itself and thereby adopting a more consistent idealism. He says as much in the second letter which I quoted above. In the "Second Introduction" to the

Wissenschaftslehre he publicly acknowledges his indebtedness to Jacobi on this subject, though he adds the twist that Kant himself never actually believed in the thing in itself:

> . . . the discovery that Kant knows nothing of any somewhat distinct from the I is anything but new. For ten years everyone has been able to see in print the most thorough and complete proof of it. It is to be found in Jacobi's [*David Hume*] . . . Jacobi has there cited and gathered together the most decisive and palpably evident of Kant's statements on this point, in the latter's own words. I have no need to do again what has been done already and could not well be done better, and refer the reader more gladly to the work in question, since the whole book, like all of Jacobi's philosophical writings, could assuredly even now be profitable reading for the present generation.[12]

So this denial of the thing in itself is one important point of agreement between Jacobi and Fichte. Fichte eliminates it as the source of sensation and thus maintains that the absolute I produces even the manifold of sensation out of itself.

The second point of influence concerns Jacobi's discussion of the relationship of the conditioned to the unconditioned (and the related topics of explanation and demonstration).[13] He claims that we explain the origin of a conditioned being insofar as we provide a mechanical explanation of its origin. And since the chain of finite conditioned beings extends back infinitely in time, there is always a conditioned being to serve as the explanation of any other conditioned being.

But what of the chain itself? Does it have a condition for its existence? Jacobi says that it must, for any representation of the conditioned must be accompanied by a representation of the unconditioned.[14] This leads him to claim that there must be an unconditioned cause of the conditioned chain of finite conditioned beings. A problem arises here, however. One of his main contentions in the seventh appendix is that the only form of explanation intelligible to us is mechanical, i.e., the linking up of conditioned beings as causes of other conditioned beings. (The attempt to explain the origins of things he calls "naturalism.") Yet if the chain of conditions must itself have a cause, this cause cannot be in the chain itself; thus the condition of the chain must lie outside of the chain. But if this is the case, then the unconditioned condition of the chain lies outside of the realm of mechanical explanation, and thus outside of the realm of explanation altogether:

. . . as long as we understand, we remain in a chain of *conditioned conditions*. Where this chain ceases, there we cease to understand, and there too ceases the coherent system, which we call nature.[15]

Yet the unconditioned chain must have an origin, but this origin must be supernatural and hence unintelligible to us:

The natural realm or the universe can arise and be brought forth from the supernatural realm in no other but *a supernatural way*.[16]

We may, if we are so inclined, call this supernatural origin "God," but this does not make the origin of the realm of nature any more intelligible to us. Since the attribution of will and intellect to an unconditioned, infinite being, who bears little if any similarity to conditioned beings who have both will and intellect, is impossible, we must face the fact that the origin of the universe, the transition from the infinite to the finite, is inexplicable.[17]

The reasoning just outlined is designed to support Jacobi's philosophy of faith. Jacobi believes that we are compelled to believe in God, but he maintains that this compulsion receives no support from philosophical demonstration. Since there is no way to demonstrate the nature of the being who created the universe—nor would the manner of this creation be intelligible, even if we could demonstrate this being's existence and attributes—the best that we can do is to have faith in a God capable of creating the universe. But this faith is neither arbitrary nor whimsical. It rests on immediate intuition in the form of religious experience. Ordinary religious experiences serve as the grounds for a traditional conception of God's nature and relation to His creation. Jacobi calls them "feelings" (*Gefühle*), indicating thereby that they are as immediate as our ordinary perceptual experiences. We no more employ them in inferences to demonstrate that God exists than we employ perceptual experiences to demonstrate that physical objects exist. Both types of experience ground our beliefs in their respective objects (assuming, of course, that we have both types, since many of us have never had religious experiences).[18] Thus faith circumvents the boundaries of mechanical explanation, but only to a certain degree, because how God creates the universe is still no clearer to us. We must simply accept that he is the creator of all things.

Jacobi's reflections on the relationship of the conditioned and the unconditioned seem to have influenced Fichte's thinking about the absolute I. The "Second Introduction" to the *Wissenschaftslehre* contains a discussion of mechanism and demonstration explicitly drawn from Jacobi, although Fichte does not refer to any particular passages in Jacobi's writings.[19] With regard to demonstration Fichte writes the following:

> The main source of all these critics' errors (the chief one being their inability to grasp that in order for the *Wissenschaftslehre* to provide a deduction of ordinary experience it must resort to something indemonstrable and thus immediately intuited) may well consist in this, that they have never attained a really clear conception of what *proof* is, and have therefore failed to realize that all demonstration is based on something absolutely indemonstrable. On this too they could have learnt from *Jacobi*, by whom this point, like so many others of which they are equally ignorant, has been fully brought to light.—Demonstration achieves only a conditioned, mediate certainty; a thing becomes certain thereby, if something else be certain. If doubt arises as to the certainty of this latter, it must be linked to the certainty of some third thing, and so on continually.[20]

While he does not explicitly refer to the seventh appendix, it seems quite probable that he had it in mind. But other passages from the first edition of *Über die Lehre des Spinoza* (1785) make the same point about demonstration. For example, in an appendix to the first edition Jacobi summarizes his position in six principles; and the fifth principle says, "Every proof presupposes something already unproven, the principle of which is revelation [*Offenbarung*]."[21] To speak of revelation is to speak of immediate intuition, whether it be the revelation of God to us—the more traditional sense of the term—or the revelation of physical things to us through sense perception. In either case demonstration is absent.

Fichte's discussions of the absolute I become clearer in light of Jacobi's influence. First, we see the origin of his insistence that the *Wissenschaftslehre* must begin with a free act of reflection on the self-positing I. This is the only way to intuit the first principle capable of providing a deduction of ordinary experience. In the "Second Introduction" Fichte uses the language of mechanism (most likely drawn from the seventh appendix) in explaining why this free act of reflection is necessary. Mechanism, by which he means the realm of nature, cannot apprehend itself, he says, and so we must appeal to free consciousness, i.e., the unconditioned realm of the absolute I, to deduce the conditioned realm of nature.[22] This act of reflection is, of course, a form of immediate intuition.

Furthermore, we can make better sense of many things which strike us as logically puzzling. Consider, for example, the following passage from part 3 of the *Wissenschaftslehre*:

The absolute I is absolutely identical with itself: everything therein is one and the same I, and belongs . . . to one and the same I; nothing therein is distinguishable, nothing manifold; the I is everything and nothing, since it is nothing *for itself*, and can distinguish no positing and no posited within itself.—In virtue of its nature it *strives* . . . to maintain itself in this condition.—There emerges in it a disparity, and hence something alien to itself. (*That* this happens can in no sense be proved a priori, but everyone can confirm it only in his own experience. Moreover, we are so far unable to say anything further at all of this alien element, save that it is *not* derivable from the inner nature of the I, for in that case it would simply not be anything distinguishable.)[23]

The context of this passage indicates that this so-called alien element is the not-I. Rather than deny that the absolute I posits a not-I within itself, Fichte maintains that it does, in spite of the fact that the I is supposed to posit itself absolutely.

One might think that there would be no logical room, so to speak, for the not-I within the absolute I, and Fichte seems to agree. In the *Wissenschaftslehre* he devotes a great deal of attention to the issue of the positing of the not-I within the absolute I, trying to remove what appears to be the contradiction of an absolute I in opposition to itself. This opposition is necessary for explaining various features of our experience and action, but it seems to violate the principle of noncontradiction. Hence Fichte's many efforts to resolve the contradiction and his many references to the circle which understanding cannot escape:

In uttering the word *explain* we are already in the realm of finitude; for all *explanation*, that is, not immediate comprehension, but a progression from one thing to the next, is a finite affair, and limitation or determination is simply the bridge we traverse to it, and which the I possesses in itself. . . . As we can also put it, therefore: the ultimate ground of all consciousness is an interaction of the I with itself, by way of a not-I that has to be regarded from different points of view. This is the circle from which the finite spirit cannot escape, and cannot wish to escape, unless it is to disown reason and demand its own annihilation.[24]

Fichte tolerates this circle, despite its logical difficulties, because he finds it within himself when he turns his gaze inward in the free act of self-reflection with which he begins the *Wissenschaftslehre*.

The influence of Jacobi here is undeniable. Fichte has taken to heart the claim that the scholar's greatest merit is to unveil and reveal existence, even if in the process we uncover logical difficulties of the kind just mentioned. Some things are not to be explained but simply accepted for what they are. Jacobi claimed to have discovered a conflict between reason and faith, saying that faith contradicted what reasoned proved to be true. Yet he accepted the deliverances of faith. Similarly, Fichte claims to have discovered a conflict between the self-positing activity of the absolute I and the explanatory ability of finite understandings. Yet he accepts the self-positing activity of the absolute I and the logical difficulties that accompany it.

At this point I would like to turn to *Allwill* because Fichte's admiration for this work ties in with the points just made about demonstration and explanation. At first glance it seems a little bizarre for Fichte to claim to see similarities between his views and those put forward in *Allwill*. Jacobi's book is, after all, a collection of letters depicting the lives of a particular stratum of the eighteenth-century German bourgeoisie.[25] Jacobi gives us sentimental scenes of domestic bliss and devotion, discussions of motherhood, and the like. Throughout the letters there also floats the somewhat shadowy figure of Eduard Allwill, an aesthete modelled on the early Goethe, whom Jacobi had befriended around the time of the original edition of the work.[26] Allwill is an attractive yet dangerous and mercurial personality, for he leads a life of great emotion and little principle. He is not a bad man, really, but the potential for great evil is there. Since geniuses do not subscribe to the rules which the rest of us abide by, the possibility of doing evil is always there. This is the chief source of the suspicion directed towards him throughout the letters.

None of the foregoing seems to have anything to do with the issues which concern Fichte, and so his enthusiasm for the book is puzzling. But if we set aside what little plot and action there are in the book and look instead at the type of people Jacobi describes, Fichte's remarks begin to make more sense. Jacobi presents himself as the editor of the collection, and in the preface added to the work in 1792 he describes himself as follows:

> Thus already as a boy the man was an enthusiast, a dreamer, a mystic. . . . All of his most important convictions rested on immediate intuition—his proofs and refutations, on facts which (as it seemed to him) were partly insufficiently noticed, partly insufficiently compared as of yet.[27]

In the introduction added to *Über die Lehre des Spinoza* in 1819 Jacobi said that *Allwill* contained the "key to my works" and referred to the pages containing the autobiographical passage just quoted.[28] Although Fichte could not have read this claim about *Allwill*, the pages referred to in the 1792 introduction stand out and draw the reader's attention. They set the tone for the letters to follow, especially since there Jacobi resolves to portray "makind as it is, explicable or inexplicable."[29]

The letters depict various characters whose most important convictions rest on immediate intuition. There is not a great deal of philosophical speculation in the letters, but there are plenty of intuitions about various matters. This is not say that the characters are irrational or benighted: they simply refuse to explain the inexplicable and thus rely on immediate intuition whenever it is appropriate to do so. We might say that they are embodiments of Jacobi's philosophy of faith. The fifteenth letter, the philosophically most interesting one, recounts a heated discussion of skepticism about the external world (in the context of a discussion of Berkeley). Allwill appears and takes the side of knowledge against the skeptic, claiming that we have an "original instinct" for the truth.[30] We know many things without recourse to first principles and inferences. They are immediately intuited as true: for example, our knowledge of the things around us in the external world.

In his thinking about faith Jacobi applies this claim about immediate intuition to the perception of God. Just as we immediately perceive the world around us, says Jacobi, we perceive God immediately. This reliance on immediate intuition obviously appealed to Fichte, and is surely one source of his interest in *Allwill*. (Its convincing portrayal of the ordinary standpoint is probably another reason for his fascination with this work.) Furthermore, this accounts for the similarities which he sees between Jacobi's book and his ideas. *Allwill* depicts people who rely on immediate intuition. They are mostly commonsense realists about the external world, while Fichte is a transcendental idealist who is also an empirical realist. Thus he believes that he can ally himself with Jacobi, for he thinks that his idealism is compatible with Jacobi's realism. Fichte routinely claims that his philosophy is compatible with common sense, and thus he believes that an alliance with realists of all sorts, especially Jacobi, is possible, as long as he precisely delineates the boundaries of idealism and realism. So he employs speculative philosophy to support the commonsense realism of the non-philosopher. And in doing so Fichte also relies on immediate intuition: that is, the free act of reflection with which the *Wissenschaftslehre* begins.

But, as we know, Jacobi repudiated the *Wissenschaftslehre* in his letter to Fichte from 1799.[31] And he did so for the same reason that he repudiated Spinoza some twenty years earlier: he refused to recognize Fichte's absolute I as an adequate conception of God. Once again, Jacobi demanded a transcendent cause of the universe, not an immanent cause of things. In his letter he does not personally accuse Fichte of atheism, but he maintains that the *Wissenschaftslehre* is atheistic:

> That it knows nothing of God would be no reproach to transcendental philosophy, since it is generally acknowledged: God cannot be *known*, but only *believed in* [*geglaubt*]. A God who could be known would be no God at all. But a *merely artificial* faith [*Glaube*] in Him is also an *impossible faith*; for insofar as it wants to be merely artificial—or merely scientific or purely speculative—it negates *natural faith*, and with that, itself as *faith*; hence all of theism.[32]

Fichte does not provide us with a transcendent God. Furthermore, an absolutely self-positing I is not to be confused with a living, personal creator of the universe. The accusation against Spinoza has simply been repeated in a different context.

In response to Jacobi's letter Fichte maintained that he had been egregiously misunderstood, mostly because Jacobi had overlooked his distinction between the ordinary and the transcendental viewpoints.[33] This distinction is mentioned in the the letter of August 30, 1795, but Jacobi seems not to have considered it in his reply to Fichte. It had already been discussed in the writings of the atheism controversy. This controversy, the spiritual heir to the pantheism controversy, was ignited by Fichte's essay "On the Basis of Our Belief in a Divine Governance of the World" (from November of 1798).[34] In it he employs this distinction to explain why ordinary religious believers (like the characters of Jacobi's *Allwill*) claim that the world is God's creation and is governed according to the laws of morality. That is, why do they believe that providence is at work in the world?

Fichte responds by saying that they do not believe in providence in virtue of the ordinary standpoint (which in this essay is identified with natural science, but which could also be construed as the commonsense point of view, given that elsewhere it is often called the "standpoint of life"). From this point of view the sensible world is a self-regulating whole, operating according to the laws of the natural sciences. Morality plays no part in this viewpoint. Only from the standpoint of speculation—that is, from Fichte's transcendental philosophy—can we understand the origin of the belief in providence. From this point of view we discover ourselves to be free beings who pursue a necessary goal of reason, and there is a moral world order that guarantees that this goal will be realized:

The sensible world proceeds peacefully along its own path, in accordance with its own eternal laws, in order to constitute a sphere for freedom. But it exercises not the least influence upon morality or immorality, and it has no power at all over a free being. Autonomous and independent, the latter soars above all nature. The goal of reason can be actualized only through the efficacious acting of a free being; moreover, in accordance with a higher law, this goal will surely be achieved through such acting. It is possible to do what is right, and thanks to this higher law, every situation is arranged for this purpose. In consequence of this same arrangement, an ethical act infallibly succeeds and an ethical one infallibly fails.[35]

But what is the relationship between this moral world order and God? Did God create it? No, says Fichte, it was not created by God. It *is* God. He succinctly formulated this idea in what was probably the most inflammatory statement of the controversy: "This living and efficaciously acting moral order is itself God. We require no other God, nor can we grasp any other."[36]

It is hardly surprising that when the time came for Jacobi to respond, he chose to characterize the *Wissenschaftslehre* as atheistic. The God described in Fichte's letter of August 30, 1795, and in the "Divine Governance" essay in no way resembles the God he worshiped. While Fichte had not repeated the letter of Spinozism, he had, according to Jacobi, stayed within its spirit. Fichte's speculative standpoint defends a conception of God that was incompatible with the one revealed to Jacobi. And since he was forced to choose between them, Jacobi chose the latter and thus rejected the philosophical point of view which denied his ordinary religious experience. His religious experience was, for him, more compelling than the speculation which contradicted it. The choice was an easy one.

After the publication of Jacobi's letter Fichte responded to the charge of atheism in several writings. Some of these never appeared in Fichte's lifetime,[37] but *The Vocation of Man* was published in 1800, shortly after the end of the atheism controversy. There he once again employs the distinction between the ordinary and speculative standpoints. Since this work lies beyond the scope of this paper, I shall simply say the following about it. Jacobi is nowhere mentioned in it, but his presence is felt throughout the book.[38] Fichte is obviously responding to what he took to be the injustices of the open letter of 1799. What Jacobi would have said, had he chosen to respond, should be clear by now. In the section on faith Fichte presents, once again, a conception of God which Jacobi could not accept, what with its language about the infinite will, the One, and so on. No living, personal God is to be found in these pages. Yet another rejection would have ensued.

So Fichte's great hopes were dashed by a man whom he admired and respected. He should have realized that Jacobi would not consent to an alliance with the *Wissenschaftslehre*. Yet he had some reason to think it possible: the influence of Jacobi on his work is apparent to anyone familiar with Jacobi's writings. Thus Jacobi was disposed to look favorably on Fichte, which he did. But he could not bring himself to accept Fichte's views on the absolute I: to him it was just an idealistic version of Spinozism, and thus yet another philosophical denial of a transcendent creator of the universe.

Despite his disappointment Fichte continued to think highly of Jacobi. In a letter to Reinhold dated January 8, 1800, Fichte placed Jacobi "far above Kant" as an influence on his thought and called him "the deepest thinker of our time."[39] Somewhat surprisingly, he did not publically attack Jacobi, whose open letter he considered unfair and uninformed. We know that Fichte was extremely sensitive to criticism, and thus his refusal to retaliate against Jacobi serves as further proof of his respect and admiration for someone whom he considered a philosophical mentor.

Notes

1. I choose these dates somewhat arbitrarily, because it is not entirely obvious which of Jacobi's works Fichte read. But the significant editions of *Allwill*, a book that Fichte admired and read many times, were published in 1775–76 and 1792. (See n. 26 for further information on the editions of *Allwill*.) Those of Jacobi's other works which might have influenced Fichte's *Wissenschaftslehre* were published between these dates. (Strictly speaking, the *Foundations* is not the *Wissenschaftslehre*, i.e., Fichte's system as a whole. Since, however, this is the book of Fichte's which most concerns me in this paper, for the sake of brevity I shall henceforth refer to it as the *Wissenschaftslehre*.)

2. F. H. Jacobi, *Auserlesener Briefwechsel* (Leipzig: Gerhard Fleischer, 1825–1827; Bern: Herbert Lang, 1969), vol. 2, pp. 183–84.

3. Ibid., pp. 207–11; Daniel Breazeale, ed. and trans., *Fichte: Early Philosophical Writings* (Ithaca, N.Y.: Cornell University Press, 1988), pp. 411–12.

4. See F. H. Jacobi, *Werke*, ed. F. Köppen and F. Roth (Leipzig: Gerhard Fleischer, 1812–1825; Darmstadt: Wissenschaftliche Buchgesellschaft, 1968), IV/1, p. 90. There Jacobi explicitly says that Lessing is a Spinozist because he does not believe "in a cause of things separate from the world."

5. Ibid., p. 72.

6. For my reading of the controversy, see chap. 1 of "Kant, Jacobi, and the Transition to Post-Kantian Idealism" (unpublished dissertation, 1993). There I argue that the basic issue of the controversy is whether or not reason can provide us with a traditional picture of God and his relation to creation. Jacobi claimed that it could not—what Spinoza offers us hardly resembles the God of the Christian tradition—and so he rejected its pronouncements in this and other matters (including that of freedom). This rejection is quite unconvincing, for it seems to be motivated simply by his distaste for Spinozism. His claims about faith seem entirely ad hoc and arbitrary. Fortunately, in his later works, especially in *David Hume*, Jacobi goes on to defend his ideas about faith. He adopts Thomas Reid's account of perception and extends it to the perception of God: that is, ordinary religious experience supports belief in God, just as ordinary visual experience supports belief in the external world. Yet Jacobi never abandons the view that the claims of Spinozism have been rationally demonstrated. So he seems to be caught between two contradictory beliefs, viz., that God exists and that God does not exist, since both possess convincing support of some sort. This tension is never resolved in his writings. (My account of Jacobi's philosophy of faith is to be found in chapter four of my dissertation.)

7. Jacobi, *Werke*, vol. 2, pp. 301–302.

8. Ibid., p. 303.

9. Ibid., p. 304.

10. Ibid., p. 307.

11. Ibid., p. 310.

12. *J. G. Fichte-Gesamtausgabe der Bayerischen Akademie der Wissenschaften*, eds. Reinhard Lauth, Hans Gliwitzky, and Erich Fuchs (Stuttgart–Bad Cannstatt: Frommann-Holzboog, 1964ff), I/4: 235; *Fichte: Science of Knowledge*, p. 54 (translation altered).

13. The seventh appendix to *Über die Lehre des Spinoza* is to be found in *Werke*, IV/2, pp. 127–62. My discussion of it concentrates on pp. 147–57.

14. Ibid., p. 152.

15. Ibid., p. 154.

16. Ibid., p. 155.

17. Ibid., pp. 156–57.

18. Jacobi explicitly denies that *Gefühle* serve as premises from which to infer conclusions: "Gefühle sind keine Erweis-Gründe. . . ." Quoted in Karl Hammacher, *Die Philosophie Friedrich Heinrich Jacobis* (Munich: Wilhelm Fink, 1969), p. 166. This quotation is taken from Jacobi's unpublished papers.

19. Fichte, *Gesamtausgabe* I/4: 258–61.

20. Fichte, *Gesamtausgabe*, I/4: 260; *Fichte: Science of Knowledge*, p. 77.

21. Jacobi, *Werke*, IV/1, p. 223.

22. Fichte, *Gesamtausgabe*, I/4: 509–10.

23. Fichte, *Gesamtausgabe*, I/2: 399–400; *Fichte: Science of Knowledge*, p. 233 (translation altered).

24. Fichte, *Gesamtausgabe*, I/2: 412–13; *Fichte: Science of Knowledge*, p. 248 (translation altered).

25. For a helpful account of the context in which Jacobi wrote *Allwill*, see Roy Pascal, "The Novels of F. H. Jacobi and Goethe's Early Classicism," *Publications of the English Goethe Society* 16 (1947): 54–89, especially pp. 54–65. Pascal sees *Allwill* and *Woldemar*, Jacobi's other novel, as investigations of the *Genie* and how he relates to the society around him. Though his interpretation ignores the philosophical issues which interested Fichte, it is useful in other respects.

26. Published in 1775–76 in *Iris* and *Der Teutsche Merkur* and included in Jacobi's *Vermischte Schriften* of 1781. An expanded and revised edition was published in 1792 and incorporated into Jacobi's *Werke* in 1812 with mostly minor changes. In his letter of August 30, 1795 Fichte does not mention which edition he had been reading, but it seems safe to assume that he would have been reading the latest one, i.e., the one from 1792. In what follows I refer to the pagination of the 1812 edition (in vol. 1 of Jacobi's *Werke*), since this is the most accessible edition for modern readers.

27. Jacobi, *Werke*, vol. 1, pp. xii–xiii.

28. Jacobi, *Werke*, IV/1, p. viii.

29. Jacobi, *Werke*, vol. 1, p. xiii.

30. Jacobi, *Werke*, vol. 1, pp. 118–23. (See pp. 145–47 for similar remarks.) The fifteenth letter is not in the 1775–76 edition of *Allwill* and thus was added after Jacobi had begun to develop his ideas on faith in more explicitly philosophical works.

31. The original letter (Fichte, *Gesamtausgabe*, III/3: 224–81) was sent to Fichte in March of 1799. A revised version was published later that year (see Jacobi, *Werke*, vol. 3, pp. 3–57). Jacobi's target in this work is somewhat unclear. He does not attack specific writings but rather what we might call the spirit of Fichte's system. And it seems reasonable to assume that he understands it based on the explanation of the *Wissenschaftslehre* offered in Fichte's letter of August 30, 1795, as well as the writings of the atheism controversy published in 1798–99.

32. Jacobi, *Werke*, vol. 3, p. 7.

33. In what follows I shall only briefly discuss this distinction. Much more can be said about it, and I recommend that the reader consult two essays by Daniel Breazeale for further information. See "The 'Standpoint of Life' and the 'Standpoint of Philosophy' in the Context of the *Jena Wissenschaftslehre* (1794–1801)," in *Transzendentalphilosophie als System: Die Auseinandersetzung zwischen 1794 und 1806*, ed. Albert Mues (Hamburg: Felix Meiner, 1989), pp. 81–104; "Philosophy and the Divided Self: On the 'Existential' and 'Scientific' Tasks of the Jena *Wissenschaftslehre*," *Fichte-Studien* 5 (1994): 117–47.

34. Fichte, *Gesamtausgabe*, I/5: 347–57.

35. Fichte, *Gesamtausgabe*, I/5: 353; Daniel Breazeale, ed. and trans., *Introductions to the Wissenschaftslehre and Other Writings* (Indianapolis: Hackett, 1994), p. 149.

36. Fichte, *Gesamtausgabe*, I/5: 354; Breazeale, *Introductions*, p. 151.

37. See the letters to Reinhold and Jacobi, both from April 22, 1799 (Fichte, *Gesamtausgabe*, III/3: 325–33, 333–37), as well as two posthumously published texts (*Johann Gottlieb Fichtes Sämmtliche Werke*, ed. I. H. Fichte [Berlin:Viet & Co., 1845–46], XI, pp. 390–94; Fichte, *Gesamtausgabe*, II/5: 190–91). The second text clearly pinpoints the disagreement: namely, the extent to which science (i.e., transcendental philosophy) can describe life. Fichte was optimistic about the powers of philosophy to do so; Jacobi was not.

38. Jacobi's influence on this work is generally recognized. See, for example, Peter Rohs, *Johann Gottlieb Fichte* (Munich: C. H. Beck, 1991), pp. 121–32, especially pp. 121–24.

39. Fichte, *Gesamtausgabe*, III/4: 180. Fichte also complains about some of the "assertions" which he considers "unworthy" of Jacobi, e.g., his insistence upon a personal God, and thus points out the chief difference between the two of them.

ON THE RECEPTION OF THE *GRUNDLAGE*

Part 5

The Early Critical Reception of the 1794 *Wissenschaftslehre*

Dale Snow

It would be a most instructive contribution to the history of the human spirit for someone to recount the reception received by various philosophical propositions upon their first appearance. It is a genuine loss that we no longer possess the first astonished judgment of contemporaries concerning some of the old systems.[1]

Composition of the *Foundations*

Fichte had suffered through a series of uncongenial tutoring positions when he received the invitation to succeed Reinhold at Jena, an offer he did not feel that he could afford to reject. However, he felt himself to be in a quandary with respect to his teaching duties, since, as he wrote to his friend Karl Böttiger:

> None of the available texts by Kant or Reinhold suits my purpose, nor can I write a textbook of my own between now and the end of next month. Thus the following expedient occurred to me: what if I were to distribute my textbook in installments during the course *as a manuscript for the use of my listeners* (since I absolutely wish to delay for a few years any presentation of my system for the public at large)? . . . In order to show that I am serious about this, the text should not be published in a regular trade edition at all, but should be distributed *only on my instructions* to my students. . . .[2]

Thus was the *Foundations* hastily conceived, and regretted at leisure. Fichte repeatedly mentions both the hurried circumstances of its composition as well as its original purpose of serving as simply an aid to students who would have the opportunity of hearing it clarified orally in trying to explain why so many misleading rumors about his views had arisen, as in this letter to Reinhold:

> Bear in mind that what you have received so far is a manuscript for the use of my students. It was hastily written while I was busy with my lectures . . . and with a thousand other very diverse activities. I had to see that the written sheets came to an end at the same time as the lectures. Of course I firmly believe that what I have *intuited*, and for the most part, what I have *thought*, is irrefutable; but part of what I have *said* may, at least in part, be quite wrong.[3]

Perhaps influenced by financial considerations, Fichte soon changed his mind about distributing the book only to his students; indeed, by April of 1794 he is writing to Böttiger that he wants "a good-sized edition . . . 2000 copies" of the *Einladungsschrift* ("On the Concept of the *Wissenschaftslehre*") printed, and "in the bookstores as soon as possible,"[4] so clearly he was intending to "invite" more than just his own students to further study of his new system. The first half became available in September, under the title *Foundations of the Entire Wissenschaftslehre: A Manuscript for the Use of his Students*. Fichte is careful to observe in the preface "I myself declare this presentation to be extremely imperfect and defective."[5] Even the advertisement for the first half that appeared in the *Allgemeine Literatur-Zeitung* mentioned its incompleteness and hasty composition, and called it "no more than a manuscript . . . for these reasons [the author] is reluctant to see it submitted to public criticism."[6]

This reluctance is all the more understandable if we take into account the severity and high standards reflected in Fichte's own early work as a reviewer, which prompted at least one friend to criticize the tone, if not the content, of his reviews of works by F. H. Gebhard and C. A. L. Creuzer. Fichte replied impatiently:

> I have . . . every right to say of total nonsense like Gebhard's, that it is total nonsense; and to correct a young man like Creuzer, who deals with Kant so inappropriately with respect to a matter which he clearly fails to understand; however, I freely admit that of all the burdens of this life, there is for me none greater than that of having to read a wide-ranging,

disorderly and unclearly written book; and then my disaffection pours out of my pen.[7]

In his polemical essay "The Annals of Philosophical Tone" of 1797 Fichte states succinctly his minimum requirements of any acceptable review of his own works:

> I will step forward and explain the reasons why I consider it philosophy's task to provide a derivation of *experience* in its entirety as the necessary condition for self-consciousness. . . . Anyone who does not agree with what I am going to say must either show that my reasons for defining philosophy in this way are insufficient, or, if he cannot do that, he must show that some of *my* specific deductions are incorrect. Or, if he can do neither of these things, he should just say nothing at all.[8]

Fichte goes on at length to paint an exceedingly sarcastic picture of the multiple injustices and stupidities of the opponents of his philosophy, of whom he speaks in the plural, although the editors of the *Gesamtausgabe* present a strong case for the conclusion that the specific occasion of this essay was G. E. Schulze's review of Fichte's *Foundations of the Philosophy of Right*.[9]

Responding to a reproach from Reinhold about the condescension in the tone of his own polemics, Fichte explains:

> You say that my tone offends and wounds persons whom it does not concern. I sincerely regret this; nevertheless, it does concern them to the extent that they do not wish to let someone tell them honestly what terrible errors they are indulging, and to the extent that they do not want to accept a bit of shame as the price for some very important instruction. Certainly the Wissenschaftslehre will have no effect upon anyone who does not value truth above all else—including his petty individual self.[10]

Without a doubt this sounds very harsh; yet I must remark in passing, although I cannot pursue it here, that one could at least make the case that Fichte demands no more of others than he does of himself, as he demonstrates in this claim in a letter to Reinhold: "I would not hesitate for a moment to give up this philosophy if I were shown that it is incorrect . . . I would believe that I had done my duty."[11]

EARLY REVIEWS

The pain of negative criticism ought to be borne, Fichte suggests above, in order that the greater good of speaking and growing closer to the truth may be served; but if he was looking for constructive criticism from his earliest critics, he was bitterly disappointed. Indeed, he was so astonished and disappointed by the initial reception of both the "invitational" work "On the Concept of the *Wissenschaftslehre*" and the 1794 *Wissenschaftslehre* that he appended two of the most negative reviews to the 1798 reprint of "On the Concept of the *Wissenschaftslehre*," in order, apparently, that the authors be hoist by their own petards (revealing in this connection is that one of the reviews is actually of Schelling's 1794 essay "On the Possibility of a Form for All Philosophy").

The editors of *J. G. Fichte in zeitgenössichen Rezensionen* reprint five reviews of "On the Concept," three of the 1794 *Wissenschaftslehre*, and three that deal with both.[12] The first to appear was a review of "On the Concept" by F. A. Weisshuhn in the *Philosophisches Journal für Moralität, Religion, und Menschenwohl*. Weisshuhn, a childhood acquaintance of Fichte's whose correspondence shows him to have been an uncritical admirer, writes an oddly patchwork review: the first part makes a diligent effort to summarize what he takes the aims and content of the work to be, but the conclusion veers off into an elliptical rejection: "I wish humbly to declare that I do not care for the scenery from the standpoint Prof. Fichte has led us to, because my ordinary eyes are shown things that they cannot grasp, doing things that the ordinary understanding cannot fathom."[13]

Some reviewers faulted Fichte's style of presentation, like the anonymous reviewer for the *Philosophisches Journal*, who hoped that "Herr Fichte will in the future be luckier with respect to the clarity and distinctness of his presentation; perhaps he has been seduced by a desire to seem new and different . . ."[14] Others criticized both the style ("An exact and reliable report of the content of this work is impossible for us, for although we have taken the greatest pains, we have not been able to clearly see its interconnections, nor discover the incomprehensibly deep meaning of its author"[15]) and a specific problem. A recurring question concerned the nature of the I as first principle:

> then how the I that posits itself as determined through the not-I all at once produces all theoretical knowledge out of itself, that is, we don't mind admitting, for us still an impenetrable mystery, even if the author often assures us that he has presented it clearly and distinctly.[16]

J. S. Beck's review, one of the ones Fichte reprints, also follows the pattern of declining to discuss the argument as a whole: "We will emphasize only a few things; for to give an overview of the whole is impossible."[17] Furthermore, and worse, he simply mocks what he does not understand. A favorite target is Fichte's use of circular arguments, a criticism repeated in later reviews:

> The reviewer has . . . found nothing so entertaining as the numerous circular arguments [*Schlusszirkel*], in which the author often finds himself, and which he each time forthrightly admits [to exist]. The reason for this forthrightness lies in the fact that these circles are not the ordinary sort. They are magic circles, that despite their circularity contain great argumentative power and glorious conclusions.[18]

The review of Schelling's essay provides more examples of the same attempts at wit: "He now begins to divide even the simplest concepts into such tiny pieces that every honest man would go blind and deaf over it, and even the subtlest scholastics must stand back in awe. At long last the author finds, after proceeding along this thorny path, the most interesting supreme principle of all philosophy, which is: *I am I*."[19]

These and other negative reactions of the Kantians were briefly alluded to in the 1798 preface to "On the Concept of the Wissenschaftslehre." Fichte observes resignedly: "[A]s is to be expected from the Kantians, the reception given to the *Wissenschaftslehre* turned out to be much more coarse and vulgar than the one given to Kant's writings."[20] He does not seem to have wavered later from the view he expressed in 1799 to L. H. Jacob, the editor of the *Annalen der Philosophie und des philosophischen Geistes*: "The manner in which I have been treated in the *Annalen* is and will continue to be inexcusable. It will always remain a black mark in the history of philosophy."[21]

However, it is notable that even so kindly disposed a critic as Fichte's friend Karl August Böttiger, whose serious-minded and positive review focused on the significance of Fichte's idea of a first principle, still managed to reflect both praise for the ambition and scope of the effort and ambivalence about the success of the presentation: "All these investigations are firmly mutually grounded in one another. However a judgment on the entire grand undertaking can only be made upon the much-desired completion [of the system]."[22]

These reactions preceded that of Kant himself by several years, although Fichte had sent Kant the first part of the *Wissenschaftslehre* in

October of 1794 and asked, in the accompanying letter, if Kant could spare the time from his many labors and have the kindness to render a judgment on it.[23] Kant did not reply until the end of 1797, and his letter excused the tardiness of his reply with the infirmities of age, and a growing disinclination to engage in the subtleties of theoretical speculation.

Yet a challenge in the name of the public that appeared in the *Erlanger Literaturzeitung* did provide the occasion for Kant to take a public stand on the *Wissenschaftslehre*.[24] In August of 1799, in the *Allgemeine Literatur-Zeitung*, Kant wrote: "In response to the formal challenge presented to me in the name of the public . . . I hereby declare: that I hold Fichte's Wissenschaftslehre to be a completely untenable system. For pure science of knowledge is nothing more or less than mere *logic*. . . . "[25] Kant goes on to insist that he is not going to take any position on the worth or meaning of Fichte's theoretical principles, but that he does resent the effrontery of those who have attributed to him the aim of merely providing a propadeutic to transcendental philosophy, and not the system of this philosophy. The critical philosophy is in no need of improvement, expansion, or transformation and will remain for all time of irreplaceable importance to the highest purposes of mankind.[26]

One might think that such a repudiation would be hard for Fichte to accept; yet as he explains to Schelling, there are two reason why he is not unduly concerned: "You ask what I have to say about Kant's declaration on my system? You can find in my papers left in Jena the very correspondence he refers to in his article. . . . "[27] Fichte goes on to point out the references in Kant's letter to his "infirmities of age" and his disinclination to engage in theoretical speculation, admissions he was unlikely to want to make publicly. Moreover, Kant's unwillingness to accept the *Wissenschaftslehre* is part of what amounts to the natural order of things:

> It is to be expected, dear Schelling, that just as the defenders of pre-Kantian metaphysics have not yet stopped saying of Kant that he has devoted himself to fruitless hair-splitting, Kant says the same to us: it is to be expected that just as they insist to Kant that their metaphysics still stands undamaged, unimprovable, and unalterable for all time, Kant says the same of his own [system] to us. Who knows where already the young brilliant mind may be working, who will attempt to go beyond the principles of the *Wissenschaftslehre*, and will attempt to prove its errors and incompleteness.[28]

REINHOLD

Busy and preoccupied as he was with his teaching, editing and writing projects, not to mention his efforts to disband the disorderly student fraternities, Fichte still seems to have noticed the considerable delay in serious critical response from those who he had eagerly expected to be most enthusiastic about his discoveries. By August of 1795 he is writing quite pointedly to Reinhold: "No verdict concerning the WL can be more important to its author than the verdict of the author of the Elementary Philosophy, who was responsible for the final step which leads to the discovery of the WL."[29] In this and other letters to Reinhold, Fichte reflects on the reasons why his writings have been misunderstood. He identifies two problems that might be called internal, that is, peculiarities of his own approach to philosophy: "My mind is so constructed that it must grasp the whole either all at once or not at all, and this explains the faulty construction of my writings."[30] A related problem is rhetorical and concerns his difficulties in fathoming the state of mind of an audience he is unfamiliar with: "The reason that I have had such bad luck as an author is that I am so little capable of placing myself in the frame of mind of the reading public; I always assume that many things are self-evident that hardly anyone else finds to be self-evident."[31]

However, there are all but insurmountable external difficulties militating against the acceptance of the *Wissenschaftslehre* as well, Fichte explains after Reinhold had finally written of his enthusiastic embrace of the *Wissenschaftslehre*: "it stands as a complete whole, grounded upon itself—the pure presentation of self-disclosing reason, the mirror for our better selves—there before the eye of my spirit."[32] Fichte admits that only at this point, after Reinhold has conclusively demonstrated his understanding, is he able to reveal that he had feared Reinhold would never really master it:

> I do not think that any man would be so perverse as to continue to stubbornly deny the truth once he has come to recognize it. Nevertheless, until the opposite is proven to me, I am willing to credit almost anyone with enough self-love and a sufficient number of dogmatically held pre-conceived opinions to stand between himself and those truths which he himself has not discovered and which contradict his own assertions—even though he himself does not realize that this happens. . . . I did not expect from you that impartiality and attendant love of truth which allows one to find his way out of an error into which he has fallen in consequence of his own dedicated efforts. Such a love of truth is not something arbitrary; it

constitutes an integral part of an already acquired character, which I did not think you possessed. Please accept now my confession of this—along with the confession of my deep respect and admiration.[33]

Fichte was so pleased by Reinhold's conversion not out of the simple desire to win others over but because Reinhold's letters had convinced him that it was the right kind of conversion; that is, Reinhold had not simply learned to repeat the "dead letter" but had come to grasp and accept it "out of an inner need [*Bedürfnis*]."[34] The ability to recognize this inner need is the proof of a character constituted by the love of truth above all things, even self-interest and vanity, and Fichte reveals in this letter just how rare he takes this type of character to be.

Reinhold's review appeared in the *Allgemeine Literatur-Zeitung* in January of 1798. The *Wissenschaftslehre*, he writes, cannot be compared to any previous system; indeed, this "reviewer finds himself in the rare situation, that the report he is to prepare deals with a *fully new* philosophy, essentially different from all preceding ones."[35] The critical philosophy, which had first established the distinction between the theoretical and the practical use of reason, did not proceed to derive both from a common principle; this is the advance made by the *Wissenschaftslehre*:

> This pure scientific use of reason has not only not been attempted by dogmaticism, nor anticipated by skepticism; even criticism knows nothing of it. . . . With it and through it a new epoch must dawn for philosophy and all the sciences, the like of which has not yet been seen in the history of the human spirit.[36]

Although Reinhold does summarize specific arguments, and quotes passages from "On the Concept of the *Wissenschaftslehre*," thus in those respects meeting Fichte's requirements for adequate reviewing, his unstinting praise and the absence of any serious reservations starts to ring hollow near the end. He even blames himself for having difficulties with Fichte's style, one of the few respects in which Fichte admitted he could make improvements: "The confusions with which [I] often had to struggle, even during repeated rereading, arose out of the contrast between the fully new way of philosophizing unique to the Wissenschaftslehre and the way in which [I] had become accustomed to philosophize. . . . "[37] The end effect is not as different from the negative reviews as Fichte might have hoped: the reader is still left with more rhetoric than solid information on the revolutionary nature of Fichte's new direction in philosophy.

JACOBI

In September 1794 Fichte sent Jacobi the published part of the *Wissenschaftslehre*. In the enclosed letter he says: "If there is any thinker in Germany with whom I wish and hope to be in agreement concerning my personal convictions, it is you, my very dear Sir—I, who from most well-known philosophical authors expect nothing but contradiction."[38] Fichte may well have been quite sincere in his high opinion of Jacobi's philosophical acumen, but he nevertheless had become wary enough of being misunderstood that he found it necessary, in a letter written a month later, to mention and debunk a popular misconception about his views: "My *Absolute I* is obviously not the *individual*, though this is how offended courtiers and irate philosophers have interpreted me, in order that they may falsely attribute to me the disgraceful theory of practical egoism. Instead, *the individual must be deduced from the absolute I. . . .*"[39]

Jacobi was not unworthy of Fichte's high opinion in the sense that he certainly took time and care in formulating his opinion on the *Wissenschaftslehre*. The resulting long letter, later infamous for its introduction of the term "nihilism," was ultimately published as the "Open Letter to Fichte" in 1799. Jacobi begins with the highest praise:

> I say it at every opportunity, and am prepared to acknowledge it publicly, that I consider you the true Messiah of speculative reason. . . . And so I continue and first proclaim you, more ardently and audibly, once again among the Jews of speculative reason their king; threaten the obstinate with recognizing you as such and accepting the Baptist of Königsberg merely as your forerunner.[40]

This might seem odd, since Jacobi had long defended the vital importance of faith and was a well-known critic of what he called the pretensions of reason. Yet Jacobi insists that he and Fichte "both want, with similar seriousness and zeal, that the science of knowledge . . . become perfected; with one difference: that you want it so that the basis of all truth, as lodging in the science of knowledge, reveal itself; I, so that this basis be revealed: the true itself is necessarily present outside of it."[41]

That is, Jacobi celebrates the achievements of both Kant and Fichte because they had brought philosophy to its highest possible development, and thus made clearer than ever what reason can and cannot accomplish. Jacobi concludes: "And thus I am altogether and throughout still the same person who in the letters on Spinoza proceeded from the miracle of perception and the inscrutable mystery of freedom and ventured in this manner with a *salto mortale* not so much to prove his philosophy, but rather daringly to present his unphilosophical obstinacy to the world."[42] Jacobi had every incentive to praise Fichte's achievement because it served his own agenda of proving that systematic philosophical speculation produces a monotonous, unbreakable web of necessary connection which rigorously excluded the possibility of freedom, moral judgement, personality and indeed the idea of a personal God at all. Thus although the "open Letter" was widely read, it, like many of Jacobi's other writings, revealed more about its author than its ostensible subject.

SCHELLING

With all the hubris of youth, Schelling announced his existence and his understanding of the *Wissenschaftslehre* to Fichte in the same letter, which accompanied a copy of his 1794 essay "On the Possibility of a Form for all Philosophy": "Some things in this book still remain unclear to me, but there is much, most especially with respect to what appears to be the main point of it, that seems to me, if I am not wholly deceiving myself, quite unambiguous."[43]

Fichte remarks in a letter to Reinhold that Schelling's essay seems "entirely a commentary on my own" but well-written enough that some who had struggled with his (Fichte's) writing seem to have found Schelling's clearer and adds "I am happy that he has appeared."[44] As late as 1798 Fichte considered him to have "the same manner . . . of philosophizing."[45]

Asked by F. I. Niethammer to review the *Wissenschaftslehre* for his new *Philosophisches Journal*, Schelling responds enthusiastically in January of 1796:

> I accept your request to review Fichte's WL with all the more pleasure, since I have not yet myself had enough time to actually *study* this work. I have not even read the practical part if it yet. So your flattering judgment that I have already become completely familiar with Fichte's philosophy is a little too flattering. Still I believe that I have grasped the spirit of it in general, even if I am as yet very little acquainted with the detail and the letter of the WL.[46]

By March he is contritely admitting that "[I've] as yet accomplished very little on the review of the Fichtean work"[47] and offering to give up the assignment. In August he asks Niethammer about the change in publisher for the *Philosophisches Journal*, due to take place at the beginning of 1797, adding that as soon as these arrangements are complete "I will at last get to work on the review of the Wissenschaftslehre."[48] Finally in November he writes accepting Niethammer's offer to become a "Mitarbeiter" at the journal, and explaining the terms on which he is willing to undertake the task of writing an extended review essay, to be published in installments: namely, that it would function primarily as a sort of background to present developments in the recent history of philosophy, prominent among which were Schelling's own ideas:

> It is with particular delight that I shall assume the continuing review essay on the latest in philosophical literature. With reviews alone one will never succeed in getting through it, and given the daily worsening misery of our countless philosophical authors "literary mass executions" will soon become all but inevitable. I will gradually offer a brief survey of the most recent history of philosophy starting with Kant and leading up to the present.[49]

The real review is contained between the lines of the "Philosophical Letters on Dogmatism and Criticism" of 1795. Schelling argues against the necessity of a single first principle, a necessity defended in the *Wissenschaftslehre*.

> As long as there is any philosophy, the Critique of Pure Reason will stand alone, while each *system* will allow another directly opposed to it. . . . The Critique of Pure Reason is not corruptible by partisan individuality and consequently it is valid for all systems, while every *system* bears the stamp of individuality on the face of it, because no system can be completed other than *practically*.[50]

Thus in Schelling's view, the *Critique of Pure Reason* itself is or contains the only genuine science of knowledge (*die eigentliche Wissenschaftslehre*). Nevertheless, *knowledge* may rise to an absolute principle and it *must* do so if it is to become a *system*. But the science of knowledge cannot possibly put up one absolute principle in order to become a system (in the narrower sense of the word). It must contain, not an absolute principle, not a definite and consummate system, but the canon for all principles and systems.[51] This is reiterated in the Sixth Letter, where Schelling argues that the choice between dogmatism and criticism "depends on the freedom of spirit we have ourselves acquired."[52]

Both of the 1797 introductions to the *Wissenschaftslehre* contained responses to this and other criticisms, which Fichte acknowledges later in a letter: "Your discussion in the *Philosophisches Journal* of the two philosophers, one idealistic and the [other] realistic . . . which I immediately gently contradicted, because I saw it to be mistaken, aroused in me the suspicion that you had not fully penetrated the *Wissenschaftslehre*, but . . . I hoped that with time you would make good what you had missed."[53]

Fichte's plaintive remark to Schelling in 1801 shows that in the end he was hardly better satisfied with the level of understanding achieved by his most promising erstwhile disciple than with the uncomprehending stupidities of his bitterest enemies. It echoes a comment made a year earlier that could stand as his last word on this vexed topic:

> The Foundations of the [Entire] Wissenschaftslehre . . . is useless, at least for those who have spoken in public about it. For, as I gather from almost all public judgments of my philosophy and from the reproaches of opponents who actually want the very same thing that I do, as well as from the objections to my philosophy and from the new efforts that are devoted to philosophy, no one yet possesses any knowledge whatsoever of *what I am attempting to do*.[54]

Fichte goes on to point out that no one, after all, has been in any way forced to criticize him or indeed to be concerned with philosophy at all. We must all learn to remain silent if we are not willing to do the honest hard work philosophy requires, or if we discover that we lack the character to recognize the truth.

NOTES

1. *Johann Gottlieb Fichtes Sämmtliche Werke*, ed. I. H. Fichte (Berlin: Viet & Co., 1845–46), I, pp. 33–34. Reprinted, along with *Johann Gottlieb Fichtes nachgelassene Werke* (Bonn: Adolphus-Marcus, 1834–35), as *Fichtes Werke* (Berlin: de Gruyter, 1971).

2. *J. G. Fichte-Gesamtausgabe der Bayerischen Akademie der Wissenschaften*, eds. Reinhard Lauth, Hans Gliwitzky, and Erich Fuchs (Stuttgart–Bad Cannstatt: Frommann-Holzboog, 1964ff), III/2: 71; letter of March 1, 1794, translation in *Fichte: Foundations of Transcendental Philosophy*, ed. and trans. Daniel Breazeale (Ithaca: Cornell University Press, 1992), p. 5 n.12.

3. Letter of July 2, 1794; translated in Breazeale, *Fichte*, p. 401.

4. Fichte, *Gesamtausgabe*, III/2: 90; letter of April 2, 1794.

5. Fichte, *Sämmtliche Werke*, I, p. 87.

6. Fichte, *Gesamtausgabe*, I/2: 183.

7. Ibid., III/2: 81; letter of March 8, 1794 to Gottlieb Hufeland.

8. *Philosophisches Journal einer Gesellschaft Teutscher Gelehrten* 5, no. 1; trans. in Daniel Breazeale, ed. and trans., *Fichte: Early Philosophical Writings* (Ithaca, N.Y.: Cornell University Press, 1988), p. 347.

9. "Die scharfen Ausfälle Fichtes in den Annalen werden allerdings erst verständlichen, wenn man erkennt, dass es niemand anders, als der Helmstädter Philosophie-Professor G. E. L. Schulze ist, gegen den sie sich richten. Dies entging freilich sowohl den Zeitgenossen als auch den späteren Beurteilern. Fichte bezeichnet ihm jedoch in den Annalen so hinreichend, dass er eindeutig ausgemacht werden kann." Fichte, *Gesamtausgabe*, I/4: 286.

10. Letter of March 21, 1791; translation in Breazeale, *Fichte*, p. 417.

11. Letter of February 7, 1795 to Reinhold; translation in Breazeale, *Fichte*, p. 397–98.

12. *J. G. Fichte in zeitgenössichen Rezensionen Wissenschaftslehre*, eds. Erich Fuchs, Wilhelm G. Jacobs, and Walter Schieche (Stuttgart-Bad Cannstatt: Frommann-Holzboog, 1995).

13. Ibid., p. 252.

14. Ibid., p. 258.

15. Ibid., p. 337.

16. Ibid., p. 339.

17. Fichte, *Sämmtliche Werke*, I, p. 75.

18. Ibid., I, p. 77.

19. Ibid., I, p. 68.

20. Breazeale, *Fichte*, pp. 99–100.

21. Letter of March 4, 1799; see in Breazeale, *Fichte*, p. 424.

22. *Fichte in zeitgenössischen Rezensionen*, p. 348.

23. Fichte, *Gesamtausgabe*, I/2: 207–208; letter of October 6, 1794: "Von innigster Verehrung gegen ihren Geist durchdrungen, den ich zu ahnden glaube; des Glüks theilhaftig, Ihren persönlichen Charakter in der Nähe bewundert zu habe; wie glüklich wäre ich, wenn meine neuesten Arbeiten von Ihnen eines günstigern Bliks gewürdigt würden, als man bisher darauf geworfen!"

24. Fichte, *Gesamtausgabe*, I/2: 211; the citation is from a review of J. G. Bühles *Entwurf der Transscendental Philosophie*: "Rec. glaubt . . . im Namen des Publicums die Bitte wagen zu dürfen, das der *Lehrer der Transcendental-Philosophie sein für die Wissenschaft so interessantes Urtheil über die Wissenschaftslehre mittheile*."

25. Fichte, *Gesamtausgabe*, I/2: 211–12.

26. See ibid., I/2: 212.

27. Ibid.

28. Ibid., I/2: 213.

29. Letter of August 29, 1795; translation in Breazeale, *Fichte*, p. 406.

30. Ibid.

31. Letter of April 22, 1799; translation in Breazeale, *Fichte*, p. 428.

32. Fichte, *Gesamtausgabe*, I/2: 222; letter of February 14, 1797.

33. Letter of March 21, 1797; translation in Breazeale, *Fichte*, p. 416.

34. Fichte, *Gesamtausgabe*, I/2: 225; letter of March 21, 1797.

35. *Fichte in zeitgenössichen Rezensionen*, p. 286.

36. Ibid., p. 292–93.

37. Ibid., p. 304.

38. Fichte, *Gesamtausgabe*, III/2, p. 202; letter of September 29, 1794.

39. Letter of August 30, 1795; translation in Breazeale, *Fichte*, p. 411.

40. *Jacobi's Werke* (Leipzig: Gerhard Fleischer, 1812–1825), bd. 3, pp. 3–57. The only English translation of which I am aware is the "Open Letter to Fichte," trans. Diana Behler, in *Philosophy of German Idealism* (New York: Continuum Publishing, 1987), pp. 122, 123–24.

41. Ibid., p. 125.

42. Ibid., pp. 135–36.

43. Letter of September 26, 1794, *F. W. J. Schelling: Briefe und Dokumente*, ed. Horst Fuhrmans (Bonn: H. Bouvier Verlag, 1962), vol. 1, p. 52.

44. Letter of July 2, 1794; translation in Breazeale, *Fichte*; as Breazeale points out, it is uncertain which of Schelling's early essays this comment refers to.

45. Fichte, *Sämmtliche Werke*, I, 35.

46. Letter of January 1, 1796, in *Briefe*, I, p. 60.

47. Letter of March 23, 1796, in ibid., I, p. 66.

48. Letter of August 30, 1796, in ibid., I, p. 89.

49. Letter of November 8, 1796, in ibid., I, p. 95.

50. "Philosophical Letters on Dogmatism and Criticism" (1795), I: 304.

51. Ibid., I: 304–305.

52. Ibid., I: 308.

53. Letter of May 31, 1801, in *Briefe* II, p. 339.

54. Fichte, *Gesamtausgabe*, II/5: 438; translation in Breazeale, *Fichte*, p. 8 n. 22.

HEGEL'S EARLY REACTION TO FICHTE'S *WISSENSCHAFTSLEHRE*
The Case of the Misplaced Adjective

George Seidel

14

The Hegelian reading of Fichte has been done before. It was done by Hegel for one, in *The Difference between Fichte's and Schelling's Systems of Philosophy* (briefly: *Differenzschrift*). Such a reading was implicit in Richard Kroner's *Von Kant bis Hegel*,[1] as also, to some extent, in Nicolai Hartmann's *Die Philosophie des deutschen Idealismus*.[2] What I should like to do here is a Fichtean reading of Hegel, more specifically that of Hegel's *Phenomenology of Spirit*, indicating how Fichte appears at important points in Hegel's argument. Further, from the examination of an important passage in the *Differenzschrift*, I hope to show that it is Fichte who provides what the French call *le choc* for the Hegelian agenda. I very much agree with the remark of H. S. Harris: "Fichte seems, more than any other writer, to have had the power to irritate Hegel into plans for theoretical reconstruction."[3] Finally, taking a cue from the Hegelian dialectical hermeneutic, I would suggest that anything one must fight so hard against must represent a potent influence.

If philosophy advances by making distinctions then Hegel's *Differenzschrift*, hastily written between the middle of May and the middle of July 1801,[4] represents the attempt on Hegel's part to advance philosophy beyond what he saw as the one-sided idealisms of both Schelling and Fichte. Relative to Fichte, one may say that Hegel evinces only a "superficial knowledge" of Fichte's philosophy;[5] that the reading is "inadequate and hence, at the least, very suspect";[6] or, relative to Schelling, with friends like Hegel who needs enemies? For although in this work Hegel ostensibly sides with Schelling, criticisms of Schelling's position are only slightly below the surface; for example, when Hegel speaks of the finite as drowned (*versenkt*) in the infinite;[7] or when he insists that there can be no principle of identity without a principle of difference as well.

Reinhard Lauth is surely correct in saying that Hegel gives a Schellingesque reading of Fichte in the *Differenzschrift*, pointing out that Hegel knew the idealism of Schelling before he knew that of Fichte.[8] Such a reading is evident in Hegel's interpretation of the intellectual intuition, which he sees as the basis for Fichte's system, reading it as the absolute identity of subject and object.[9] However, this is closer to Schelling's notion of intellectual intuition than it is to that of Fichte. In the 1797 "Second Introduction" to the *Wissenschaftslehre*, para. 5, Fichte describes intellectual intuition as no more than the philosopher's awareness of its self as an acting consciousness. "It is the immediate consciousness that I act, and what I enact: it is that whereby I know something because I do it," as Fichte says in the "Second Introduction."[10] As Lauth notes, Fichte always maintained, against Schelling, that the intellectual intuition will give nothing objective.[11]

In the same way that he reads Fichte through the eyes of Schelling, Hegel reads Kant through the eyes of Fichte. The *Wissenschaftslehre*, according to Hegel—and here he agrees with Fichte—represents the completion of the Kantian philosophy put in a more logical fashion.[12] In the *Differenzschrift*, for example, Hegel maintains that the dichotomy between nature and freedom is carried over from Kant into Fichte,[13] and that both of them miss nature, whereas Schelling does not. In his *Glauben und Wissen* (1802) Hegel speaks of the incompleteness of the Absolute in Fichte, that it is deficient (*mangelhaft*). Here again Hegel conflates Kant and Fichte, taking Fichte's Absolute as though it were a Kantian totality. For while it may be true that freedom is a possible meaning for the Absolute Self in Fichte, and while freedom is an heuristic idea for Kant and, in a kingdom of good wills, a totality, the two notions cannot be identified. It is the difference between an idea and an ideal. On the same score, Hegel inquires how the in-finite, if it is a totality, can avoid calling up the finite that is within it.[14] Hegel may have written a *Differenzschrift* relative to Fichte and Schelling, he did not write one contrasting Fichte and Kant.

Hegel's criticisms of Fichte's philosophy remain constant throughout his works. He had little but disdain for Fichte's popular writings.[15] His principal criticism of Fichte throughout is that Fichte's philosophy represents a subjective or formal idealism, a one-sided, *für sich* certainty not yet come to its *anundfürsich* truth.[16] In the *Differenzschrift* Hegel characterizes the identity of subject and object in Fichte as a subjective identity, a subjective subject-object, in contradistinction to Schelling's objective subject-object identity. Hence, it is a subjective idealism.[17]

Another of Hegel's basic criticisms, obvious already in the *Differenzschrift*, is that of Fichte's *Sollen* ("Ich *soll* gleich Ich sein.").[18] It is the "I am I" of the absolutely posited Absolute Self that turns into an "I should be I"—which absolute "I" is never really to be—that gets pilloried as the bad or negative infinity in Hegel. Indeed, in Hegel's view it *never* really is, not once Fichte reaches the third fundamental principle of finite self/finite non-self reciprocity.[19] Relative to the finite self, the infinite or ideal self of Fichte is a tangentially trajected wild hair. And Hegel cannot abide such unresolved contradictions, above all one as basic as that between finite and infinite. This is why Hegel prefers Schelling's identity philosophy over that of Fichte. Fichte's philosophy cannot, in Hegel's view, resolve itself into a genuine identity.

The strategy Hegel employs in the *Differenzschrift* is to argue that Fichte's three fundamental principles represent three absolute acts of the self (*drey absolute Akte des Ich*). However, because there is a plurality of acts involved they are actually finite, "really relative," as he says, since they are acts of one and the same consciousness.[20] According to one commentator, " . . . the Fichtean system has equated Reason with pure consciousness, with the unhappy result that its guiding ideal is Reason as it is conceived in finite form."[21] What follows from this is that Fichte's Absolute is really finite.[22] According to Hegel, Fichte's *Reflexions-philosophie* sets up an absolute opposition between infinite and finite (*beide mit ihrer Entgegensetzung gleich absolut sind*), the identity of the finite and the infinite posited only conceptually in the form of the infinite.[23] It is because Fichte's positings and oppositings are abstract (separate) that his *Sollen*, in Hegel's view, goes on unto infinity, or nothing.[24] The mere negating of finitude in order to produce infinitude produces a negative infinity; it becomes an abyss of nothing (*Abgrund des Nichts*) into which everything sinks.[25]

Aside from the fact that this sounds more like Schelling's Absolute than the Absolute Self of Fichte, Hegel's attempt to finesse Fichte on this score is highly suspect. The difficulty is with that phrase *die drey absolute Akte des Ich*, the three absolute acts of the self. The adjective "absolute" is misplaced. Hegel has it modifying acts (*Akte*). It is thus that he can argue that the posited infinite or absolute self is really finite, since plural acts of consciousness are involved, rendering them relative or finite. Hegel's *suppositum* must, however, be denied. For while it is true that the first and second fundamental principles are, indeed, posited (or opposited) absolutely, the third fundamental principle, the positing of finite self and non-self in reciprocal relation to each other, clearly is not. What Hegel misses, and what a closer reading of the *Wissenschaftslehre* on his part might have revealed, is that the limit produced (the x) is the product of an act (*Handlung*), namely y, a limiting.[26] It is by no means an "absolute act." Indeed, a more careful reading of Fichte would have indicated that even the acts whereby the first and second fundamental principles are posited or opposited are, in the final analysis, acts of the finite consciousness. Precisely my point, one can hear Hegel saying, the perspective of the finite ego is inadequate to reach the infinite.[27]

Nevertheless, when Hegel argues that a finite act cannot generate an absolute and, hence, the absolute is really finite, he undercuts his own crucial distinction between understanding and reason, and hence the basis for his criticism of the bad or negative infinity, which is integral to his Kant-Fichte critique. Again, the adjective "absolute" is misplaced. For the issue is precisely whether the act that produces the fundamental principles of the *Wissenschaftslehre* is that of the finite or the absolute self. In one breath Hegel has the absolute posited by the finite self—all well and good—but only to be able to say that the infinite self is thus really finite. In an earlier breath he has the finite self being produced by an absolute act. He cannot have it both ways. The act, or acts, that produce the *Wissenschaftslehre* must be posited by the finite or by the absolute self, one or the other. Now one can say that Hegel is simply bringing his characteristic dialectic to bear upon Fichte (and Schelling). However, this will not work, as by this time Hegel had not yet fully developed his dialectical method, something he will be able to do in time for the *Phenomenology* only after he had reflected more deeply upon the procedural thinking of both Fichte and Schelling.

It is sometimes said that Hegelian dialectic swallows other philosophies whole—Fichtean Sauerkraut, Schellingian Wurst, washed down with stale Kantian beer from Krug's *Krug*. Some say Schelling gets the last burp. However, not only does Hegel sometimes trim and cut up what he consumes, I would argue that it is that indigestible misplaced adjective "absolute" in Hegel's *Differenzschrift* that provokes the extensive belch that is the *Phenomenology of Spirit*, wherein Hegel recalls the steps of finite spirit to its truth in absolute spirit, an Absolute that is and wills to be with us from the very start.[28] Now Schelling likewise insists that the Absolute must be there from the start.[29] Indeed, unless it were so given there would be no freedom or even philosophy.[30] And later Hegel will again appear to side with Schelling against Fichte, as he does in the *Differenzschrift*, when he asserts that substance had to be there if subject would come to be (*Phänomenologie*, p. 314, #439).

Fichte is seldom expressly mentioned in the *Phenomenology*. Early on there is a reference to a phrase from the essay that precipitated the atheism controversy, namely the notion of God as the "moral order of the world" (*moralische Weltordnung*).[31] Yet, in reading Hegel's 1806 work, Fichte would have been gratified to find himself appearing throughout, especially in the crucial introductory portions to the major sections, functioning as the *point de départ* for Hegel's dialectical treatment. Where and how Fichte will appear in this "Gallerie von Bildern" has been foreshadowed already in the *Differenzschrift* and that misplaced adjective. At this point it may be well to take a cursory Fichtean-guided tour through Hegel's phenomenological museum, concentrating particularly upon those introductory sections.

Fichte appears, right at the start, in the mediating ego or self behind the certainty of "Sense Certainty," or so-called immediate experience. The difficulty with such imagined "immediacy" is that it represents mediated knowledge as soon as it is reflected upon (by the self). (*Phänomenologie*, pp. 79–80, ##90 ff.) Such immediacy is reminiscent of the *Vorstellungen* accompanied by a feeling of necessity from Fichte's "First Introduction" to the *Wissenschaftslehre*. For as soon as one attempts to grasp the meaning of an "immediate" experience it is mediated, if only by the universals contained within language. In other words, Hegel's solution to the problem of sensation is not all that different from that of Fichte or Kant, as becomes clear in the following section on "Perception" (pp. 71 ff., ##111 ff.). It is the matter of a synthesis of the properties contained in the experienced "thing," a representational synthesis behind which must necessarily stand an actual (Fichte) or possible (Kant) "I think."

The section on Force and Understanding with its reciprocity hearkens back to Fichte's third fundamental principle and the variable reciprocity of activity and passivity between positing self and opposed non-self. Indeed, Hegel specifically ties up the interaction of forces with perception (pp. 84–85, #136) as the play of forces between form and content in perceptual experience. And, of course, at the end of this section there is the allusion to that charming fragment from Novalis about lifting the veil in the temple of Isis and discovering . . . one's self (p. 102, #165), the (self-)consciousness that is necessarily behind any positing or op-positing or understanding of forces. Thus, the transition from consciousness to self-consciousness.

At the beginning of the section on self-consciousness Fichte is very much out in the open, with an "in itself" and a "for itself" given both to, and in, one and the same consciousness. However, Hegel makes a point of saying that (Fichte's) absolutely posited I = I of the first fundamental principle really goes nowhere (*bewegungslose Tautologie*; p. 104, #167). Still, the basis for the development of self-consciousness is Fichtean, namely desire (". . . das Selbstbewußtsein . . . ist *Begierde*"; p. 107, #174). There is that whole panoply of words in the latter part of the *Wissenschaftslehre* associated with striving (*Streben*), namely feeling (*Gefühl*), drive (*Trieb*), longing (*Sehnen*), etc. Granted, in Fichte it is a striving toward the Absolute Self, whereas for Hegel it is the more pedestrian desire toward recognition by an-other self-consciousness. Still, that which fuels the desire, and that toward which it ultimately aims is, in both cases, the same, namely freedom (p. 111, #187). Further, it is the doubling of self-consciousness in an-other self-consciousness, a contribution from Fichte ("No I, no You; no You, no I"), that is the basis for the fight for recognition in the master-slave dialectic.[32] To be a self is to be recognized as a self by an-other self, as Fichte says in the *Naturrecht* of 1796. It is, of course, in this work that Fichte spells out the basis for the "social contract," my limiting my freedom in order to allow for the freedom of the other.[33] The basis for this position is, however, already in the *Wissenschaftslehre* of 1794. The *Naturrecht* is, as indicated in the title, is established "in accordance with the principles of the WL."

Fichte is present in Hegel's characterization, or better, caricature of Stoicism, which becomes a code word for the Fichtean notion of freedom as "abstract": one thinks one is free because one is free to think (whatever one likes). The dialectic ends with the standoff between the slave Epictetus in his chains and the master of the world, Marcus Aurelius, on his throne (p. 117, #199), neither really free. It is an empty freedom, condemned to lifelessness (*Leblosigkeit*), since it is the mere notion of freedom, not the real thing (*nicht die lebendige Freiheit selbst*; p. 118, #200).

Fichte is, of course, lumped together with the romantics, Jews, and mediaeval monastics in Hegel's treatment of the "Unhappy Consciousness" and the *unendliche Sehnsucht* after an *unerreichbare Jenseits* (pp. 125–26, #217), the eternal yearning for an unreachable beyond. The language is Fichtean as well as romantic.[34] The "beyond" is simply Fichte's Absolute Self, infinite and ideal, and essentially unattainable. This section of the *Phenomenology* contains one of Hegel's enduring criticisms of Fichte, the notion of the bad or negative infinity, that is, an infinite absolute one "should" (*soll*) forever seek but can never achieve.

Typically, the section on *Vernunft* begins with Fichte, or perhaps better, with Hegel's reading of Kant through the eyes of Fichte. Reason is consciousness' certainty of being all reality.[35] This subjective idealism is immediately criticized as one-sided and empty. Hegel questions that if for Kant the transcendental unity of apperception is the truth of knowledge (p. 136, #238), or for Fichte the absolute self ("I am I"; p. 134, #234) is the ultimate category, because it is, in the final analysis, the source of all categories (pp. 134–35, #235), then it must be the source of the category of unity as well. If so, then there must be another schema to join the synthetic unity of the transcendental ego together with the category of unity (p. 135, #236).[36]

Further, Hegel points to that other "certainty," namely the non-self. The self tries, but cannot fully rid itself of the *Ding an sich*, the unknowable, but necessary, limit to possible knowledge (in Kant) or (in Fichte) the *fremder Antoß*, the external shove or check, required for the development of consciousness (Cf. p. 136, #238). Hegel is likely referring to that section of the *Wissenschaftslehre* where Fichte outlines the necessity of the "check" for representation.[37] However, Fichte insists that this cannot be established on the basis of the theoretical, only on the basis of the practical, reason. Again, Hegel's conflation of Fichte and Kant is really unfair to both.

The Actualization of Self-Consciousness section (pp. 193 ff., ##347 ff.) opens out onto the world of society and morality. The basis for such social interaction is basically Fichtean. In Fichte there is a reciprocity between subject and object, self and non-self (the other), depending upon the activity/passivity (positing and non-positing activity) of the self[38]— again, "No I, no You; no You, no I." Similarly in Hegel, I recognize in the equally independent other the same free unity as in myself: they as I, I as them (p. 195, #351). The basis for this social interaction in Hegel, as in Fichte, is the same: the activity of freedom. The difference is Hegel's submerging of Fichte's attempted philosophical basis for Rousseau's social contract theory. Hegel does not approve of Rousseau, at least not of his notion of a primitive state of nature.

The section on Spirit (pp. 238 ff., ##438 ff.) presents the abstract (separated) ethical world of Fichte, which, given Fichte's projection of the Absolute, reminiscent of the Unhappy Consciousness, produces a moral world rent asunder into a here and a beyond (*Diesseits und Jenseits*). In morality, conscience (*Gewißheit*) is certain (*gewiß*) of itself (pp. 323 ff., ##596 ff.). However, this certainty is one-sided and purely subjective. This becomes part and parcel of Hegel's criticism of the Kantian and Fichtean notion of conscience, which, in his view, is too individualistic, since it ignores the important element of *Sittlichkeit*, the social dimension of morality. It is of a piece with Hegel's criticism of Fichte's notion of freedom as abstract and one-sided in that it ignores the notion of community (*das Gemeine*). Again, this criticism of Fichte's moral philosophy pushes back into Kant since, for Hegel, Fichte correctly interprets Kant. For Kant conscience is the consciousness which is duty for itself (*für sich selbst*).[39] This is, of course, circular, since in this tradition "for itself" is a code name for consciousness. In other words, the consciouness of duty is the consciousness of duty. In Hegel's view Kant, and Fichte, trap conscience in a totally one-sided subjectivity.

In the section on Religion (pp. 363 ff., ##672 ff.) as also in the early part of Revealed Religion (pp. 400 ff., ##748 ff.), Hegel rehearses the position of the Unhappy Consciousness and with it what he regards as the Fichtean notion of God as Absolute Self. In this treatment he often confuses Fichte with Schelling, as also with Kant, taking the more Kantian-based notion of God as the moral order of the world from the Fichte of the Atheism Controversy period.

Finally, at the very end of the *Phenomenology of Spirit* (pp. 430 ff., ## 803), Hegel does a quick review of modern philosophy starting with Descartes. He finds Fichte's Absolute Self (the I = I) to be a Self that is not substance. On the other hand, Schelling's substance is not subject, and hence not Spirit. Hegel thus ends the *Phenomenology* as he began it, playing the Fichte card against Schelling, the Schelling card against Fichte. For already in the Preface, in the passage immediately following his criticism of the (Schellingian?) notion of the Absolute as the night in which all cows are black (p. 17, #16), Hegel had insisted that the truth must be subject as well as substance. This means, further, that Hegel ends the *Phenomenology* as he began the *Differenzschrift*, with the idealisms of Fichte and Schelling as both abstract and one-sided.

Now, of course, we the philosophers, who know what is going on behind the back of the Hegelian consciousness, know full well that Hegel is not being fair to either Fichte or Schelling. He turns them into pure positions, then plays one off against the other. Still, Hegel had to write the *Phenomenology*. He had to write it especially because of Fichte and the adjective he (Hegel) had misplaced in the *Differenzschrift*. For instead of writing *die drey absolute Akte des Ich*, had he placed the adjective *absolute* in front of the *Ich*—*die Akte des absoluten Ich*—and read Fichte more carefully, he could have seen the difficulty. For then it would have been obvious that the adjective *endlichen* would be the one appropriate to the context. For then he would have recognized much sooner the need to travel the long and rocky dialectical road from finite to infinite spirit, albeit with an absolute that is and wills to be there with us from the start. Hegel had to travel this road if only to get to the distinction between the infinite of the understanding (the bad or negative infinity) and the infinite of reason (the good infinity),[40] since only then could he come to the realization that the in-finite is not the negative of the finite but, with a typically Hegelian dialectical twist, it is the finite that is the negative.[41]

It has been suggested that in the *Encyclopedia* (##413 ff.) the view of the human personality presented there, namely the "I," the "me" and the "myself," becomes the model for Hegel's doctrine of the Trinity.[42] The "me," the objective consciousness, is the Word that creates the objects of which there is consciousness, as also the Word that is rendered objective (made flesh). The "I" of self-consciousness is the Father, who knows and wills. The "myself" is the Spirit that mediates the two. If this is true, then Fichte's *für sich* Absolute Self holds the exalted place of God the Father in Hegel's Trinitarian theology. Which is curious. For it means that what Hegel has taken over from Fichte's *Wissenschaftslehre* is the I = I of the Absolute Self as God the Father, when Fichte thought his work was a Christology, that is, about God the Son.[43]

Of course, it is not the first time in the history of Western thought that a philosopher thought he was doing one thing while a later thinker decides he was doing something else.

NOTES

1. Richard Kroner, *Von Kant bis Hegel*, 2d ed. (Tübingen: Mohr, 1961), which reads the history of German philosophy from Kant on as the march toward its culmination in the philosophy of Hegel.

2. Nicolai Hartman, *Die Philosophie des deutschen Idealismus*, 2d ed., (Berlin: de Gruyter, 1960).

3. *Hegel's Development Toward the Sunlight 1770–1801* (Oxford: Clarendon, 1972), p. 252.

4. H. S. Harris, *The Difference between Fichte's and Schelling's System of Philosophy* (Albany: State University of New York Press, 1977), p. 1.

5. Helmut Girndt, "La critique de Fichte par Hegel dans la 'Differenzschrift' de 1801," *Archives de philosophie* 28 (1965) 37–61.

6. J. G. Naylor, "La controverse de Fichte et de Hegel sur l'"indifférence,'" *Archives de Philosophie* 41 (1978): 53.

7. G. W. F. Hegel, *Gesammelte Werke* (Hamburg: Meiner, 1968), vol. 4, p. 63.

8. "La position spéculative de Hegel dans son écrit 'Differenz des

Fichte'schen und Schelling'schen Systems der Philosophie' à la lumière de la Théorie de la Science," *Archives de Philosophie* 46 (1983): 70 n. 43. Lauth adds that Hegel reads Fichte from the point of view of a dogmatist ontology regarding nature (p. 101), insisting also that Hegel entirely misses the notion of freedom in the *Wissenschaftslehre* (pp. 102–103).

9. Hegel, *Gesammelte Werke*, vol. 4, p. 76. As Tom Rockmore correctly notes, the intellectual intuition in Fichte and in Schelling is different: for Schelling it is in relation to a supersensuous object; for Fichte only in relation to an activity. *Hegel's Circular Epistemology* (Bloomington: Indiana University Press, 1986), p. 57.

10. *Johann Gottlieb Fichtes Sämmtliche Werke*, ed. I. H. Fichte (Berlin: Viet & Co., 1845–46), I, 463. Reprinted, along with *Johann Gottlieb Fichtes nachgelassene Werke* (Bonn: Adolphus-Marcus, 1834–35), as *Fichtes Werke* (Berlin: de Gruyter, 1971).

11. Reinhard Lauth, in "Philosophie transcendental et idéalisme absolu," *Archives de Philosophie* 48 (1985): 376, suggests further that it is this that constitutes the difference between the transcendental philosophy and absolute idealism.

12. *Hegel's Lectures on the History of Philosophy*, trans. E. S. Haldane and F. H. Simson (London: Routledge & Kegan Paul, 1896), vol. 3, pp. 479, 481.

13. Hegel, *Gesammelte Werke*, vol. 4, pp. 59–60.

14. Ibid., vol. 4 p. 391.

15. Hegel characterizes Fichte's *Anweisung zum seligen Leben* as edifying but philosophically worthless. *Hegel's Lectures on the History of Philosophy*, vol. 3, pp. 480–81.

16. Ibid., vol. 3, p. 486.

17. Hegel, *Gesammelte Werke*, vol. 4, pp. 33–34, 63.

18. Ibid., vol. 4, p. 33.

19. According to Walter Cerf and H. S. Harris, Hegel always maintained that Fichte could not get the Absolute Self back once he had abandoned it in favor of the finite self and finite non-self of the third fundamental principle. Cf. *G. W. F. Hegel: Faith and Knowledge* (Albany: State University of New York Press, 1977), p. 173, n.41.

20. Hegel, *Gesammelte Werke*, vol. 4, p. 37.

21. Cf. G. J. Percesepe, "*Telos* in Hegel's Differenz des Fichte'schen und Schelling'schen Systems der Philosophie," *Philosophy Research Archives* 10 (1985): 399.

22. Hegel "Glauben und Wissen," in *Gesammelte Werke*, vol. 4, pp. 322, 390.

23. " . . . die Identität des Endlichen und Unendlichen selbst wieder nur in der Form des Unendlichen als Begriff gesetzt. . . . " Ibid. vol. 4, p. 346.

24. Ibid., vol. 4, pp. 391, 396.

25. Ibid., vol. 4, p. 413.

26. *J. G. Fichte's sämmtliche Werke*, ed. J. H. Fichte (Berlin: Veit, 1845), vol. 1, pp. 107–108.

27. As Tom Rockmore notes in *Hegel's Circular Epistemology*, p. 63.

28. " . . . if [the Absolute] were not with us, in and for itself, all along, and of

its own volition..." *Phänomenologie des Geistes*, eds. W. Bonsiepen and R. Heede, *Gesammelte Werke*, vol. 9, p. 73 (henceforth page numbers to this edition of the work will be followed by paragraph numbers from the A. V. Miller translation of *Hegel's Phenomenology of Spirit*, Oxford: Clarendon Press, 1977, here #73). Martin Heidegger, in his *Hegel's Concept of Experience* (trans. K. R. Dove [New York: Harper & Row, 1970], p. 149), makes a great deal of this passage from Hegel's Introduction to the *Phenomenology* as one of the bases for his critique of Hegel's "Onto-theology."

29. "Das Absolute kann nur durch das Absolute gegeben seyn . . ." *Vom Ich also Prinzip #3 (Schelling Werke*, [Stuttgart: Frommann-Holzboog, 1980], vol. 2, p. 90).

30. Ibid., vol. 2, p. 133.

31. Hegel, *Gesammelte Werke*, vol. 9, 20, #23. Hegel also alludes to Fichte's *Sonnenklarer Bericht* on p. 38, #51.

32. Cf. H. S. Harris, "*Fichtes Verdienst*," *Revue internationale de Philosophie* 49 (1995): 86.

33. Fichte, *Sämmtliche Werke*, III, 51–52.

34. "Der Mensch soll sich der an sich unerreichbaren Freiheit ins Unendliche immer mehr nähern." Fichte, *Sämmtliche Werke*, I, 117.

35. With the section on Reason, according to H. S. Harris ("*Fichtes Verdienst*," p. 87), "Fichte is now—very properly—given credit for the organization of the Critical Philosophy into a 'formal idealism.' "

36. Though Cyril O'Regan is surely correct in pointing out a difference between Kant and Fichte on this score. Kant, in Fichte's view, did not think through the relation between the categories and the transcendental ego. Fichte did. Hegel, indeed, picks up on Fichte here and assigns the categories to Reason. *The Heterodox Hegel* (Albany: State University of New York Press, 1994), p. 87.

37. Fichte, *Sämmtliche Werke*, I, 217 ff.

38. Fichte, *Sämmtliche Werke*, I, 183, 218.

39. In part 4, section 4 of *Die Religion innerhalb der Grenzen der blossen Vernunft*.

40. A point he had not fully reached in the *Differenzschrift* according to Joseph Naylor: "The relation between the understanding and 'reason' remains arbitrary and unexplained" in the Hegel of the *Differenzschrift*, p. 65.

41. H. S. Harris makes essentially the same point when he says, "Fichte's Ego-theory was the key which enabled Hegel to 'invert' all of our ordinary comon-sense assumptions about the relation of the eternal and the temporal world." "*Fichtes Verdienst*," p. 91.

42. *Hegel's Philosophy of Mind: Being Part Three of the Encyclopaedia of the Philosophical Sciences* (1830), trans. W. Wallace, together with the *Zusätze* in Boumann's Text (1845), trans. A. V. Miller (Oxford: Clarendon, 1971), pp. 153ff.

43. Cf. my "The Atheism Controversy of 1799 and the Christology of Fichte's *Anweisung zum seligen Leben* of 1806," in *New Perspectives on Fichte*, eds. T. Rockmore and D. Breazeale, (Amherst, N.Y.: Humanity Books, 1995), pp. 146–49.

CONTRIBUTORS

MICHAEL BAUR is assistant professor of philosophy at Fordham University in New York, where he teaches courses on German idealism and the philosophy of law, and secretary of the Hegel Society of America and of the American Catholic Philosophical Association. He has published articles on various nineteenth- and twentieth-century thinkers, including Kant, Fichte, Hegel, Heidegger, and Gadamer, and has recently translated Fichte's *Foundations of the Natural Right*. He has also coedited *Hegel and the Tradition* and *The Emergence of German Idealism*.

CURTIS BOWMAN is a lecturer in the philosophy department at the University of Pennsylvania and the author of articles on Kant, Fichte, Jacobi, Husserl, and Derrida. He is also a contributor of translations to Immanuel Kant's *Notes and Fragments* (forthcoming).

DANIEL BREAZEALE is professor of philosophy at the University of Kentucky and a specialist in the history of German philosophy from Kant to Nietzsche. He is the cofounder of the North American Fichte Society and the coeditor, with Tom Rockmore, of several volumes of essays on Fichte. He has also translated and edited three volumes of Fichte's writings and has published dozens of essays on Fichte and lectured widely on his philosophy.

ARNOLD FARR is assistant professor of philosophy at St. Joseph's University in Philadelphia. His areas of research include German idealism, critical theory, and philosophy of race. He most recently coedited *Marginalization* *and Mainstream America* (2000) and is currently completing a book entitled *Reading Farrakhan/Reading America: An Essay of Racialized Consciousness and the Obstruction of Understanding.*

STEVEN HOELTZEL is assistant professor of philosophy at James Madison University. He is the author of several essays on Fichte and the early Schelling, and is currently working on a book-length study of the origins and early development of Schelling's philosophy.

PIERRE KERSZBERG is professor of philosophy at the University of Toulouse in France, specializing in Kant and German idealism, phenomenology, and the history and philosophy of science. He has published many articles and books, including his most recent, *Kant et la Nature* (1999).

C. JEFFERY KINLAW is professor of philosophy at McMurry University in Abilene, Texas. He is the author of several essays and coeditor of *Reading Schleiermacher: Essays in Honor of Michael Ryan.*

TOM ROCKMORE is professor of philosophy at Duquesne University and has edited or coedited several books, including *On Hegel's Epistemology and Contemporary Philosophy* (1995); *Heidegger, German Idealism, and Neo-Kantianism* (2000); and *New Essays on the Precritical Kant* (2001).

GEORGE SEIDEL is professor of philosophy at St. Martin's College in Lacey, Washington. He is the author of *Fichte's Wissenschaftslehre of 1794: A Commentary on Par I* (1993), *Knowledge as a Sexual Metaphor* (2000), and *Toward a Hermeneutics of Spirit* (2000).

DALE SNOW is an associate professor of philosophy at Loyola College in Maryland. She is the author of *Schelling and the End of Idealism* (1996) and articles on Kant, Fichte, Schelling, and Schopenhauer. She has also published translations of Jacobi, Dilthey, Schelling, and Karl-Otto Apel.

JERE PAUL SURBER is professor of philosophy and cultural studies at the University of Denver, and vice president/program chair of the Hegel Society of America. He has published extensively on most of the major figures of German idealism, especially on their views of language and its significance for their broader systematic projects.

MICHAEL G. VATER is associate professor of philosophy at Marquette University in Milwaukee, Wisconsin. He edited and translated F. W. J. Schelling's *Bruno* and is author of various articles on Fichte, Schelling, and Hegel. His teaching interests include environmental philosophy and Buddhist metaphysics.

VLADIMIR ZEMAN is professor of philosophy at Concordia University in Montreal. The coeditor and coauthor of *Transcendental Philosophy and Everyday Experience* (1997), he has published primarily on the history of philosophy, with a concentration in German philosophy, and on cybernetics and system theory.

GÜNTER ZÖLLER is professor of philosophy and chair of the Department of Philosophy at the University of Munich, and president of the International J. G. Fichte Society. He has authored or edited several books, including *Minds, Ideas, and Objects: Essays on the Theory of Representation in Modern Philosophy* (1993) and *Figuring the Self: Subject, Individual, and Others in Classical German Philosophy* (1997), and he is currently editing *The Cambridge Companion to Fichte*, Kant's *Prolegomena to Any Future Metaphysics*, and *Fichte's System of Ethics*. He has published more than sixty articles on Kant and German idealism.